MARINE ECOLOGY AND ITS LAWS

Encyclopaedia of Environmental Pollution and Awareness in 21st Century Series

MARINE ECOLOGY AND ITS LAWS

Edited by

V.K. Prabhakar

ANMOL PUBLICATIONS PVT. LTD.

NEW DELHI - 110 002 (INDIA)

ANMOL PUBLICATIONS PVT. LTD.
4374/4B, Ansari Road, Daryaganj
New Delhi - 110 002

Marine Ecology and Its Laws

First Edition, 2001
ISBN 81-261-0915-7

PRINTED IN INDIA

Published by J.L. Kumar for Anmol Publications Pvt. Ltd., New Delhi - 110 002 and Printed at Mehra Offset Press, Delhi.

CONTENTS

PREFACE

Today, the cry of 'pollution' is heard from all the nooks and corners of the world, and pollution has become a major threat to the very existence of mankind on this earth. It is the major challenge of our times.

The pollution of various resources has gone to such an extent that we are unable to breathe fresh air and drink fresh water. On one hand the advancements of science and technology have added to the human comforts by giving us automobiles, electrical appliances, supersonic jets, space crafts, better medicines, better chemicals to control harmful insects and other pests, etc., but, on the other hand, they have given us a very serious problem to face that is pollution. Pollution can be defined as an undesirable change in the physical, chemical, or biological characteristics of the air, water, or land that can harmfully affect health, survival, or activities of humans or other living organisms.

Pollution does not have to cause physical harm, but it may merely interfere with human activities. Chemicals spewed into the air or water from an industrial plant may be harmful to humans and other organisms living near the plant. If expensive pollution controls are required, the plant may be forced to shut down.

India faces serious problems of environmental pollution. Water pollution from untreated community and industrial wastes flowing into our rivers and streams, increasing use of fertilizers and pesticides, dumping of organic wastes etc., are issues that need to be tackled on the priority. Air pollution, long regarded as the bane of industrialised nations looms large over our country.

Here, we are presenting this *Encyclopaedia of Environmental Pollution and Awareness in Twenty-first Century* in fifty volumes with a view to make people conscious about the seriousness of pollution in the present time and its remedies.

Editor

1 Oslo Convention for the Prevention of Marine Pollution by Dumping from Ships and Aircraft (1972)

The Contracting Parties

Recognizing that the marine environment and the living resources which it supports are of vital importance to all nations;

Mindful that the ecological equilibrium and the legitimate uses of the sea are increasingly threatened by pollution;

Recognizing that converted action by Governments at national, regional and global levels is essential to prevent and combat marine pollution;

Noting hat this pollution has many sources, including dumping from ships and aircraft and discharges through ricers, estuaries, outfalls and pipelines Within national jurisdiction, that it is important that states use the best practicable means to prevent such pollution, and that products and processes which will minimize the amount of harmful waste requiring disposal should be developed;

Being Convinced that international action to control the pollution of the sea by the dumping of harmful substances from ships and aircraft can and should be 'taken without delay, but that this action should not discussion of measures to control other sources of marine pollution as soon as possible;

Considering that the states bordering the North-East Atlantic have a particular responsibility to protect the of this region;

Have agreed as follows:

Article 1

The Contracting Parties pledge themselves to take all Possible steps to prevent the pollution of the sea by substances that are liable to create hazards to human health, to harm living resources and marine life, to damage amenities or to interfere with other legitimate uses of the sea.

Article 2

The area to which this Convention applies shall be the high seas and the territorial sea which are situated

(a) within those parts of the Atlantic and Artic Oceans and their dependent seas which lie north or 36° north latitude and between 42° west longitude and 51° east longitude, but excluding

   (i) the Baltic Sea and Belts -lying to the south and east of lines drawn from Hasenore Head to Gniben Point, from Korshage to Spodsbierg and from Gilbierg Head to the Kullen, and

   (ii) the Mediterranean Sea and its dependent seas as far as the point of intersection of the parallel of 36° north latitude and the meridian of 5°36 west longitude.

(b) within that part of the Atlantic Ocean north latitude and between 44° west longitude and 42° west longitude.

Article 3

The Contracting Parties agree to apply the measures which they adopt in such a way as to prevent the diversion or dumping of harmful substances into seas outside the area to which this Convention applies.

Article 4

The Contracting Parties shall harmonize their policies and introduce, individually and in common, measures to Prevent the pollution of the sea by dumping by or from ship and aircraft.

Article 5

The dumping of the substances listed in Annex I to this Convention is prohibited.

Article 6

No waste containing such quantities of the substances and materials listed in Annex II to this Convention as the Commission established under the provisions of Article 16, hereinafter referred to as "the Commission", Shall define as significant, shall be dumped without a specific permit in each case from the appropriate national authority or authorities. When such permits are issued, the provisions of Annexes II and III to this Convention shall be applied.

Article 7

No substance or material shall be dumped without the approval of the appropriate national authority or authorities. When such approval is granted, the provisions of Annex III to this Convention shall be applied.

Article 8

1. The provisions of Article 5,6 and 7 shall not apply in case of *force majeure* due to stress of weather or any other cause when the safety of human life or of a ship or aircraft is threatened.. Such dumping shall immediately be reported to the Commission, together with full details of the circumstances and of the nature and quantities of the substances and materials dumped.
2. The provisions of Article 5 shall not apply where these substances occur as trace contaminants in waste to which they have not been added for the purpose of being dumped. However, such dumping shall remain subject to Article 6 and 7.

Article 9

If a Contracting party in an emergency considers that a substance listed in Annex (1) to this Convention cannot be disposed of an land without unacceptable danger or damage, the contracting Party concerned shall forthwith consult the Commission. The Commission shall recommend methods of storage or the most satisfactory means of destruction or disposal under the prevailing circumstances. The Contracting Party shall inform the Commission of the steps adopted in pursuance of its recommendation.

Article 10

The composition of the waste shall be ascertained by the appropriate national authority or authorities in accordance with the provisions of Annex III to this Convention before any permit or approval for the dumping of waste at sea is issued.

Article 11

Each Contracting Party shall keep, and transmit to the Commission, according to a -standard procedure, record of the nature and the quantities of the substances and materials dumped under permits or approvals issued by that Contracting Party, and of the date, places and methods of dumping.

Article 12

Contracting Parties agree to establish complementary or joint programmes of scientific and technical research, including research on alternative methods of disposal of harmful substances, and to transmit to each other the information so obtained. In doing so they will have regard to the work carried out by the appropriate international organizations and agencies.

Article 13

The Contracting Parties agree to institute, in cooperation with appropriate international organizations and agencies, complementary or joint programmes for monitoring the distribution and effects of pollutants in the area to which this Convention applies..

Article 14

The Contracting Parties pledge themselves to promote, with the competent specialized agencies and other international bodies, measures concerning the protection of the marine environment against pollution caused by oil and oily wastes other noxious or hazardous cargoes, and radioactive materials.

Article 15

Each Contracting Party undertakes to ensure compliance with the provisions of this Convention:

(a) by ships and aircraft registered in its territory;

(b) by ships and aircraft loading in its territory the substances and materials which are to be dumped;

(c) by ships and aircraft believed to be engaged in dumping within its territorial sea.

2. Each Contracting Party undertakes to issue instruction to its maritime inspection vessels and aircraft and to other appropriate services to report to its authorities any, incidents or conditions on the high seas which give rise to suspicions that dumping in contravention of the provisions of the present Convention has occurred or is about to occur.
3. Each Contracting Party shall take in its territory appropriate measures to prevent and punish conduct in contravention to the provisions of this Convention.
4. The Contracting Parties undertake to assist one another as appropriate in dealing with pollution incidents involving dumping at sea, and to exchange information on methods of dealing with such incidents.
5. The Contracting Parties further agree to work together in the development of cooperative procedures for the application of the Convention, particularly on the high seas.
6. Nothing in this Convention shall abridge sovereign immunity to which certain vessels are entitled under international law.

Article 16

A Commission, made up of representatives of each of the Contracting Parties, is hereby established. The Commission shall meet at regular intervals and at any time when, due to special circumstances, it is so decided in accordance with the Rules of Procedure.

It shall be the duty of the Commission

(a) To exercise overall supervision over the implementation of this Convention;

(b) To receive and consider the records of permits and

approvals issued and of dumping which has taken place, as provided for in Articles 8, 9 and 11 of this Convention, and to define the standard procedure to be adopted for this purpose;

(c) To review generally the condition of the seas within the area to which this Convention applies, the efficacy of the control measures being adopted, and the need for any additional or different measures;

(d) To keep under review the contents of the Annexes to this Convention, and to recommend such amendments, additions or deletions as may be agreed;

Article 18

1. The Commission shall draw up its own Rules of Procedure which shall be adopted by unanimous vote. The Government of Norway shall call the first meeting of the Commission as soon as practicable after the coming into force of this Convention.

2. Recommendations for modification of the Annexes to this Convention in accordance with Article 17 (d) shall be adopted by a unanimous vote in the Commission, and the modifications contained therein shall enter into fore after unanimous approval by the Governments of the Contracting Parties.

Article 19

For the purpose of this Convention:

1. "Dumping" means any deliberate disposal of substances and materials into the sea by or from ship or aircraft other than:

   (a) any discharge incidental to or derived from the normal operation of ships and aircraft and their equipment;

(b) the placing of substances and materials for a purpose other than the mere disposal thereof, if not contrary to the aim of this Convention.

2. "Ships and aircraft" means sea-going vessels and air-borne craft of any type whatsoever. This expression includes air-cushion craft, floating craft whether self-propelled or not, and fixed or floating platforms.

Article 20

This Convention shall be open for signature at Oslo until 15th August 1972 by the States invited to participate in the Conference on Marine Pollution, held there from 19th to 22nd October, 1971.

Article 21

This Convention shall be subject to ratification. The instruments of ratification shall be deposited with the Government of Norway.

Article 22

This Convention shall be open for accession by any state referred to in Article 20. The Contracting Parties may unanimously invite other States to accede to the Convention. The instruments of accession shall be deposited with the Government of Norway.

Article 23

1. This Convention shall enter into force on the thirtieth day following the date of deposit of the seventh instrument of ratification or accession.

2. For each State ratifying or acceding to the Convention after the deposit of the seventh instrument of ratification or accession, the Convention shall enter into force on the thirtieth day after deposit by such State of its instrument of ratification or accession.

Article 24

At any time after two years from the date on which this Convention has come into force with respect to a Contracting Party, that Party may withdraw from the Convention by means of a notice in writing addressed to the depository Government. Any such withdrawal shall take effect twelve months after the date of its receipt.

Article 25

A conference for the purpose of revising or amending this Convention may be convened by the depositary Government at the request of the Commission adopted by a two-thirds majority.

Article 26

The depositary Government shall inform the Contracting Parties and the States referred to in Article 20:

(a) of signatures to this Convention, of the deposit of instruments of ratification or accession, and of the receipt of a notice of withdrawal, in accordance with Articles 20, 21, 22 and 24;

(b) of the date on-which this Convention will come into force in accordance with Article 23;

(c) of the receipt of notification of approval relating to modifications of the Annexes to this Convention and of the entry into force of such modifications 'in accordance with Article 18.

Article 27

The original of this Convention, of which the English and French texts are equally authentic, shall be deposited with the Government of Norway, which shall send certified copies thereof to the Contracting Parties and to the State,, referred to in Article 20, and which shall transmit *E* certified copy to the Secretary-General of the United Nations for registration and publication in accordance with Article 102 of the Charter of the United Nations.

In witness whereof the undersigned, being duly authorized by their respective Governments have signed this Convention.

Done at Oslo, this fifteenth day of February 1972.

Annex I

The following substances are listed for the purposes of Article 5 of the Convention:

1. Organohalogen compounds and compounds which may form such substances in the marine environment, excluding those which are on-toxic, or which are rapidly converted in the sea into substances which are biologically harmless;
2. Organosilicon compounds and compounds which may form such substances in the marine environment, excluding those which are non-toxic, or which are rapidly converted in the sea into substances which are biologically harmless;
3. Substances which have been agreed between the Contracting Parties as likely to be carcinogenic under the conditions of disposal;
4. Mercury and mercury compounds;
5. Cadmium and cadmium compounds;
6. Persistent plastics and other persistent synthetic materials which may float or remain in suspension in the sea, and which may seriously interfere with fishing or navigation, reduce amenities, or interfere with other legitimate uses of the sea.

Annex II

1. The following substances and materials requiring special care are listed for the purposes of Article 6:

(a) Arsenic, lead, copper, zinc, and their compounds, cyanides and fluorides, and other bulky wastes which may present a serious obstacle to fishing or navigation;

(b) Substances which, though of a non-toxic nature, may become harmful due to the quantities in which they are dumped, or which are liable to seriously reduce amenities.

2. The substances and materials listed under paragraph 1(c) above should always be deposited in deep water.

3. In the issuance of permits or approvals for the dumping of large quantities of acids and alkalies, consideration should be given to the possible presence in such wastes of the substances listed in paragraph 1 above.

4. When, in the application of the provisions of Annexes 11 and III, it is considered necessary to deposit waste in deep water, this should be done only when the following two conditions are both fulfilled;

   (a) that the depth is not less than 2000 metres,

   (b) that the distance from the nearest land is not less than 150 nautical miles.

Annex III

Provisions governing the issue of permits and approval for the dumping of wastes at sea.

1. Characteristics of the waste

   (a) Amount and composition;

   (b) Amount of substances and materials to be deposited per day (per week, per month);

   (c) Form in which it is presented for dumping, i.e. whether as a solid, sludge or liquid; .

(d) Physical, chemical, biochemical and biological properties;

(e) Toxicity;

(f) Persistence;

(g) Accumulation in biological materials or sediments;

(h) Chemical and physical changes of the waste after release, including possible formation of new compounds;

(i) Probability of production of taints reducing marketability of resources (fish, shellfish, etc.)

2. Characteristics of dumping site and method of deposit

(a) Geographical position, depth and distance from coast;

(b) Location in relation to living resources in adult or juvenile phases;

(c) Location in relation to amenity areas;

(d) Methods of packing, if any;

(e) Initial dilution achieved by proposed method of release;

(t) Dispersal, horizontal transport and vertical mixing characteristics;

(g) Existence and effects of current and previous discharges and dumping in the area.

3. General considerations and conditions

(a) Interference with shipping, fishing, recreation, mineral, extraction, desalination, fish and shellfish culture, areas of special scientific importance and other legitimate use of the sea;

(b) In applying these principles the practical availability of alternative means of disposal or elimination will be taken into consideration.

2 Convention on the Prevention of Marine Pollution by Dumping of Wastes and Other Matter (1972)

Preamble

The Contracting Parties to this Convention

Recognizing that the marine environment and the living organisms which it supports are of vital importance to humanity, and all people have an interest in assuring that it is so managed that its quality and resources are not impaired;

Recognizing that the capacity of the sea to assimilate wastes and render them harmless, and its ability to regenerate natural resources, is not unlimited;

Recognizing that States have, in accordance with the Charter of the United Nations and the principles of international law, the sovereign right to exploit their own resources pursuant to their own environmental policies, and the responsibility to ensure that activities within their jurisdiction or control do not cause damage to the environment of other States or of areas beyond he limits of national jurisdiction;

Recalling Resolution 2749 (XXV) of the General Assembly of the United Nations on the principles governing the sea bed and the ocean floor and the subsoil thereof, beyond the limits of national jurisdiction;

Nothing that marine pollution originates in many sources, such as dumping and discharges through the atmosphere, rivers, estuaries, outfalls and pipelines, and that it is important that States use the best practicable means to prevent such pollution and develop products and processes which will reduce the amount of harmful wastes to be disposed of;

Being convinced that international action to control the pollution of the sea by dumping can and must be taken without delay but that this action should not preclude discussion of measures to control other sources of marine pollution as soon as possible and;

Wishing to improve protection of the marine environment by encouraging States with a common interest in particular geographical areas to enter into appropriate agreements supplementary to this Convention:

Have agreed as follows:

Article I

Contracting Parties shall individually and collectively promote the effective control of all sources of pollution of the marine environment, and pledge themselves especially to take all practicable steps to prevent the pollution of the sea by the dumping of waste and other matter that is liable to create hazards to human health, to harm living resources and marine life, to damage amenities or to interfere with other legitimate uses of the sea.

Article II

Contracting Parties shall, as provided for in the following Articles, take effective measures individually, according to their scientific, technical and economic capabilities, and collectively, to prevent marine pollution caused by dumping and shall harmonize their policies in this regard.

Article III

For the Purposes of this Convention:

1. (a) "Dumping" means:

    (i) any deliberate disposal at sea of wastes or other matter from vessels, aircraft, platforms or other man-made structures at sea;

    (ii) any deliberate disposal at sea of vessels, aircraft, platforms or other man-made structures at sea;

    (b) "Dumping" does not include:

    (i) the disposal at sea of wastes or other matter incidental to, or derived from the normal operations of vessels, aircraft, platforms or other man-made structures at sea and their equipment, other than wastes or other matter transported by or to vessels, aircraft, platforms or other man-made structures at sea, operating for the purpose of disposal of such matter or derived from the treatment of such wastes or other matter on such vessels, aircraft platforms or structures;

    (ii) placement of matter for a purpose other than the mere disposal thereof, provided that such placement is not contrary to the aims of this convention;

    (c) The disposal of wastes or other matter directly arising from, or related to the exploration, exploitation and associated off-shore processing of seabed mineral resources will not be covered by the provisions of this Convention.

2. "Vessels and aircraft" means waterborne or airborne craft of any type whatsoever. This expression includes air cushioned craft and floating craft, whether self-propelled or not.

3. "Sea" means all marine waters other than the internal waters of States.

4. "Wastes or other matter" means material and substance of any kind, form or description.

5. "Special permit" means permission granted specifically on application in advance and in accordance with Annex II and Annex III.

6. "General permit" means permission granted in advance and in accordance with Annex III.

7. "The Organisation" means the organisation designated by the Contracting Parties in accordance with Article XIV2.

Article IV

1. In accordance with the provisions of this Convention Contracting Parties shall prohibit the dumping of any wastes or other matter in whatever form or condition except as other wise specified below:-

   (a) The dumping of wastes or other matter listed in Annex I is prohibited;

   (b) The dumping of wastes or other matter listed in Annex II requires a prior special permit;

   (c) The dumping of all other wastes or matter required a prior general permit.

2. Any permit shall be issued only after careful consideration of all the factors set forth in Annex III, including prior studies of the characteristics of the dumping site; as set forth in Section B and C of that Annex.

3. No provision of this Convention is-to be interpreted as preventing a Contracting Party from prohibiting, insofar as that Party is concerned, the dumping of wastes or other matter not mentioned in Annex 1. That Party shall notify such measures to the Organisation.

Article V

1. The provisions of Article IV shall not apply when it is necessary to secure the safety of human life or of vessels, aircraft, platforms or other man-made structures at sea in cases of force caused by stress of weather, or in any case which constitutes a danger to human life or a real threat to vessels, aircraft, platforms or other man-made structures at sea, if dumping appears to be the only way of averting the threat and if there is every probability that the damage consequent upon such dumping will be less than would otherwise occur. Such dumping shall be so conducted as to minimize the likelihood of damage to human or marine life and shall be reported forthwith to the Organisation.

2. A Contracting Party may issue a special permit as an exception to Article IV 1a, in emergencies, posing unacceptable risk relating to human health and admitting no other feasible solution. Before doing so the Party shall consult any other country or countries that are likely to be affected and the Organization which, after consulting other Parties, and international organizations as appropriate, shall, in accordance with Article XIV4e promptly recommend to the Party the most appropriate procedures to adopt. The Party shall follow these recommendations to the maximum extent feasible with the time within which action must be taken and with the general obligation to avoid damage to

the marine environment and shall inform the Organisation of the action it takes. The Parties pledge themselves to assist one another in such situations.

3. Any contracting Party may waive its rights under Paragraph 2 at the time of, or subsequent to ratification of, or accession to this Convention.

Article VI

1. Each Contracting Party shall designate an appropriate authority or authorities to:
   (a) issue special permits which shall be required prior to, and for, the dumping of matter listed in Annex II and in the circumstances provided for in Article V2;
   (b) issue general permits which shall be required prior to and for the dumping of all other matter;
   (c) keep records of the nature and quantities of all matter permitted to be dumped and the location, time and method of dumping;
   (d) monitor individually, or in collaboration with other Parties and competent international organisations, the condition of the sea for the purposes of this Convention.
2. The appropriate authority or authorities of a Contracting Party shall issue prior special or general permits in accordance with paragraph 1 in respect of matter intended for dumping:
   (a) loaded in its territory;
   (b) loaded by a vessel or aircraft registered in its territory or flying its flag, when the loading occurs in the territory of a State not party to this Convention.

3. In issuing permits under sub-paragraphs 1a and b above, the appropriate authority or authorities shall comply with Annex III, together with such additional criteria, measures and requirements as they may consider relevant.

4. Each Contracting Party, directly or through a Secretariat established under a regional agreement, shall report to the organisation, and where appropriate to other Parties, the information specified in sub-paragraphs c and d of paragraph 1 above, and the criteria, measures and requirements it adopts in accordance with paragraph 3 above. The procedure to be followed and the nature of such reports shall be agreed by the Parties in consultation.

Article VII

1. Each Contracting Party shall apply the measures required to implement the present Convention to all:
   (a) vessels and aircraft registered in its territory or flying its flag;
   (b) vessels and aircraft loading in its territory or territorial seas matter which is to be dumped;
   (c) vessels and aircraft and fixed or floating platforms under its jurisdiction believed to be engaged in dumping.

2. Each Party shall take in its territory appropriate measures to prevent and punish conduct in contravention of the provisions of this Convention.

3. The Parties agree to cooperate in the development of procedures for the effective application of this Convention particularly on the high seas, including procedures for the reporting of vessels and aircraft observed dumping in contravention of the Convention.

4. This Convention shall not apply to those vessels and aircraft entitled to sovereign immunity under international law. However each party shall ensure by the adoption of appropriate measures that such vessels and aircraft owned or operated by it act in a manner consistent with the object and purpose of this Convention, and shall inform the Organisation accordingly.

5. Nothing in this Convention shall affect the right of each Party to adopt other measures, in accordance with the principles of international law, to prevent dumping at sea.

Article VIII

In order to further the objectives of this Convention, the Contracting Parties with common interests to protect the marine environment in a given geographical area shall endeavour, taking into account characteristic regional features, to enter into regional agreements consistent with this Convention for the prevention of pollution, especially by dumping. The Contracting Parties to the present Convention shall endeavour to act consistently with the objectives and provisions of such regional agreements, which shall be notified to them by the Organisation. Contracting Parties shall seek to cooperate with the Parties to regional agreements in order to develop harmonized procedures to be followed by Contracting Parties to the different conventions concerned. Special attention shall be given to cooperation in the field of monitoring and scientific research.

Article IX

The Contracting Parties shall promote, through collaboration within the Organisation and other international bodies, support for those Parties which request it for:

(a) the training of scientific and technical personnel;

(b) the supply of necessary equipment and facilities for research and monitoring;

(c) the disposal and treatment of waste and other measures to prevent or mitigate pollution caused by dumping; preferably within the countries concerned, so furthering the aims and purposes of this Convention.

Article X

In accordance with the principles of international law regarding State responsibility for damage to the environment of other State or to any other area of the environment, caused by dumping of wastes and other matter of all kinds, the Contracting Parties undertake to develop procedures for the assessment of liability and the settlement of disputes regarding dumping.

Article XI

The Contracting Parties shall at their first consultative meeting consider procedures for the settlement of disputes concerning the interpretation and application of this Convention.

Article XII

The Contracting Parties pledge themselves to promote, within the competent specialised agencies and other international bodies, measures to protect the marine environment against pollution caused by:

(a) hydrocarbons, including oil, and their wastes;

(b) other noxious or hazardous matter transported by vessels for purposes other than dumping;

(c) wastes generated in the course of operation of vessels, aircraft, platforms and other man-made structures at sea;

(d) radioactive pollutants from all sources, including vessels;

(e) agents of chemical and biological warfare; wastes or other matter directly arising from, or related to the exploration, exploitation and associated off-shore processing of seabed mineral resources.

The Parties will also promote, within the appropriate international organisation, the codification of signals to be used by vessels engaged in dumping.

Article XIII

Nothing in this Convention shall prejudice the Codification and development of the law of the sea by the United Nations Conference on the law of the Sea convened pursuant to Resolution 2750C (XXV) of the General Assembly of the United Nations nor the present or future claims and legal views of any State concerning the law of the sea and the nature and extent of coastal and flag state jurisdiction. The Contracting Parties agree to consult at a meeting to be convened by the Organisation after the Law of the Sea Conference, and in any case not later than 1976, with a view to defining the nature and extent of the right and the responsibility of a coastal state to apply the Convention in a zone adjacent to its coast.

Article XIV

1. The Government of the United Kingdom of Great Britain and Northern Ireland as a depositary shall call a meeting of the Contracting Parties not later than three months after the entry into force of this Convention to decide on organisational matters.
2. The Contracting Parties shall designate a competent organisation existing at the time of that meeting to be responsible, for secretariat duties in relation to

this Convention. Any Party to this Convention not being a member of this Organisation shall make an appropriate contribution to the expenses incurred by the Organisation in performing these duties.

3. The Secretariat duties of the Organisation shall include:

   (a) the convening of consultative meetings of the Contracting Parties not less frequently than once every two years and of special meetings of the Parties at any time on the request of two-thirds of the Parties;

   (b) preparing and assisting, in consultation with the Contracting Parties and appropriate International Organisations, in the development and implementation of procedures referred to in sub-paragraph 4e of this Article;

   (c) considering enquiries by, and information from the Contracting Parties, consulting with them and with the appropriate International Organisations, and providing recommendations to the Parties on questions related to, but not specifically covered by the Convention;

   (d) conveying to the Parties concerned all notifications received by the Organisation in accordance with Articles IV 3, V 1 and 2, VI 4, XIII, XV, XX and XXI.

Prior to the designation of the Organisation these functions shall, as necessary, be performed by the depositary, who for this purpose shall be the Government of the United Kingdom, of Great Britain and Northern Ireland.

4. Consultative or special meetings of the Contracting Parties shall keep under continuing review the

implementation of this Convention and may, inter alia:

(a) review and adopt amendments to this Convention and its Annexes in accordance with Article XV;

(b) invite the appropriate scientific body or-bodies to collaborate with and to advise the Parties or the Organisation on any scientific or technical aspect relevant to this Convention, including particularly the content of the Annexes:

(c) receive and consider reports made Pursuant to Article VI 4;

(d) promote cooperation with and between regional organisations concerned with the prevention of marine pollution;

(e) develop or adopt, in consultation with appropriate International Organisations, procedures referred to in Article V 2, including basic criteria for determining exceptional and emergency situations, and procedures for consultative advice and the safe disposal of matter in such circumstances, including the designation of appropriate dumping areas, and recommend accordingly;

(f) consider any additional action that may be required.

5. The Contracting Parties at their first consultative meeting shall establish rules of procedure as necessary.

Article XV

1. (a) At meetings of the Contracting Parties called in accordance with Article XIV amendments to this

Convention may be adopted by a two-thirds majority of those present. An amendment shall enter into force for the Parties which have accepted it on the sixtieth day after two-thirds of the Parties shall have deposited an instrument of acceptance of the amendment with the Organization. Thereafter the amendment shall enter into force for any other Party 30 days after that Party deposits its instrument of acceptance of the amendment.

(b) The Organization shall inform all Contracting Parties of any requests made for a special meeting under Article XIV and of any amendments adopted at meetings of the Parties and of the date on which each such amendment enters into force for each Party.

2. Amendments to the Annexes will be based on scientific or technical considerations. Amendments to the Annexes approved by a two-thirds majority of those present at a meeting called in accordance with Article XIV shall enter into force for each Contracting Party immediately on notification of its acceptance to the Organization and 100 days after approval by the meeting for all other Parties except for those which before the end of the 100 days make a declaration that they are not able to accept the amendment at that time. Parties should endeavour to signify their acceptance of an amendment to the Organization as soon as possible after approval at a meeting. A Party may at any time substitute an acceptance for a previous declaration of objection and the amendment previously objected to shall thereupon enter into force for that Party.

3. An acceptance or declaration of objection under this Article shall be made by the deposit of an

instrument with the Organization. The Organization shall notify all Contracting Parties of the receipt of such instruments.

4. Prior to the designation of the Organization, the Secretarial functions herein attributed to it, shall be performed temporarily by the Government of the United kingdom of Great Britain and Northern Ireland, as one of the depositaries of this Convention.

Article XVI

This Convention shall be open for signature by any State at London, Mexico city, Moscow and Washington from 29 December 1972 until 31 December 1973.

Article XVII

This Convention shall be subject to ratification. The instruments of ratification shall be deposited with the Governments of Mexico, Union of Soviet Socialist Republics, United Kingdom of Great Britain and Northern Ireland, and the United States of America.

Article XVIII

After 31 December 1973, this Convention shall be open for accession by any State. The instruments of accession shall be deposited with the Governments of Mexico, Union of Soviet Socialist Republics, United Kingdom of Great Britain and Northern Ireland, and the United States of America.

Article XIX

1. This Convention shall enter into force on the thirtieth day following the date of deposit of the fifteenth instrument of ratification or accession.

2. For each Contracting Party ratifying or acceding to the Convention after the deposit of the fifteenth

instrument of ratification or accession, the Convention shall enter into force on the thirtieth day after deposit by such Party of its instrument of ratification or accession.

Article XX

The depositaries shall inform Contracting Parties:

(a) of signatures to this Convention and of the deposit of instruments of ratification accession or withdrawal, in accordance with Articles XVI, XVII, XVIII and XXI and

(b) of the date on which this Convention will enter into force, in accordance with Article XIX.

Article XXI

Any Contracting Party may withdraw from this Convention by giving six months' notice in writing to a depositary, which shall promptly inform all Parties of such notice.

Article XXII

The original of this Convention of which the English, French, Russian and Spanish texts are equally authentic, shall be deposited with the Governments of Mexico, the Union of Soviet Socialist Republics, the United kingdom of Great Britain and Northern Ireland, and the United States of America who shall send certified copies thereof to all States.

Annex I

1. Organohalogen compounds.
2. Mercury and mercury compounds.
3. Cadmium and cadmium compounds.
4. Persistent plastics and other persistent synthetic materials, for example, netting and ropes, which

may float or may remain in suspension in the sea in such a manner as to interfere materially with fishing, navigation, or other legitimate uses of the sea.

5. Crude oil, fuel oil, heavy diesel oil, and lubricating oils, hydraulic fluids, and any mixtures containing any of these, taken on board for the purpose of dumping.

6. High-level radioactive wastes or other high-level radioactive matter, defined on public health, biological or other grounds, by the competent international body in this field, at present the International Atomic Energy Agency, as unsuitable for dumping at sea.

7. Materials in whatever form (e.g. solids, liquids, semi-liquids, gases or in a living state) produced for biological and chemical warfare.

8. The preceding paragraphs of this Annex do not apply to substances which are rapidly rendered harmless by physical, chemical or biological processes in the sea provided they, do, not:

   (i) make edible marine organisms unpalatable, or

   (ii) endanger human health or that of domestic animals.

   The consultative procedure provided for under Article XIV should be followed by a Party if there is doubt about the harmlessness of the substance.

9. This Annex does not apply to wastes or other materials (e.g. sewage sludges and dredged spoils) containing the matters referred to in paragraph 1-5 above as trace contaminants. Such wastes shall be subject to the provisions of Annexes II and III as appropriate.

Annex II

The following substances and materials requiring special care are listed for the purposes of Article VI 1a.

A. Wastes containing significant amounts of the matters listed below:

arsenic

lead)

copper)

zinc) and their compounds

organosilicon compounds)

cyanides

fluorides

pesticides and their by-products not covered in Annex 1.

B. In the issue of permits for the dumping of large quantities of acids and alkalis, consideration shall be given to the possible presence in such wastes of the substances listed in paragraph A and to the following additional substances:

beryllium)

chromium) and their compounds

nickel)

vanadium)

C. Containers, scrap metal and other bulky wastes liable to sink to the sea bottom which may present a serious obstacle to fishing or navigation.

D. Radioactive wastes or other radioactive matter not included in Annex I. In the issue of permits for the dumping of this matter, the Contracting Parties should take full account of the recommendations of the competent international body in this field, at present the International Atomic Energy Agency.

Annex III

Provisions to be considered in establishing criteria governing the issue of permits for the dumping of matter at sea, taking into account Article IV 2 include:

A. *Characteristics and Composition of the matter*

1. Total amount and average composition of matter dumped (e.g., per year).
2. Form (e.g., solid, sludge, liquid, or gaseous).
3. Properties: physical (e.g., solubility and density), chemical and biochemical (e.g., oxygen demand, nutrients) and biological (e.g., presence of viruses bacteria, yeasts, parasites),
4. Toxicity.
5. Persistence: physical, chemical and biological.
6. Accumulation and biotransformation in biological materials or sediments.
7. Susceptibility to physical, chemical and biochemical changes and interaction in the environment with other dissolved organic and inorganic materials.
8. Probability of production of taints or other changes reducing market ability of resources (fish, shellfish, etc.).

B. *Characteristics of Dumping Site and Method of Deposit*

1. Location (e.g., co-ordinates of the dumping area, depth and distance from the coast), location in relation to other area (e.g., amenity areas, spawning, nursery and fishing areas and exploitable resources).
2. Rate of disposal per specific period (e.g., quantity per day, per week, per month).

3. Methods of packaging and containment, if any.

4. Initial dilution achieved by proposed method of release.

5. Dispersal characteristics (e.g., effects of currents, tides and wind on horizontal transport and vertical mixing).

6. Water characteristics (e.g., temperature, pH, salinity, stratification, oxygen indices of pollution dissolved oxygen (DO), chemical oxygen demand (COD), biochemical oxygen demand (BOD)—nitrogen present in organic and mineral form including ammonia, suspended matter, other nutrients and productivity).

7. Bottom characteristics (e.g., topography, geochemical and geological characteristics and biological productivity).

8. Existence and effects of other dumpings which have been made in the dumping area (e.g., heavy metal background reading and organic carbon content).

9. In issuing a permit for dumping, Contracting Parties should consider whether an adequate Scientific basis exists for assessing the consequences of such dumping, as outlined in this Annex, taking into account seasonal variations.

C. *General Considerations and Conditions*

1. Possible effects on amenities (e.g., presence of floating or stranded material, turbidity, objectionable odour, discolouration and foaming).

2. Possible effects on marine life, fish and shellfish culture, fish stocks and fisheries, seaweed harvesting and culture.

3. Possible effects on other uses of the sea (e.g., impairment of water quality for industrial use, underwater corrosion of structures, interference with ship operations from floating materials, interference with fishing or navigation through deposit of waste or solid objects on the sea floor and protection of areas of special importance for scientific or conservation purposes).

4. The practical availability of alternative land-based methods of treatment, disposal or elimination, or of treatment to render the matter less harmful for dumping at sea.

3 Convention on the Prevention of Pollution by Ships (1973)

The Parties to the Convention

Being Conscious of the need to preserve the human environment in general and the marine environment in particular,

Recognizing that deliberate, negligent or accidental release of oil and other harmful substances from ships constitutes a serious source of pollution,

Recognizing also the importance of the International Convention for the Prevention of Pollution of the Sea by Oil, 1954, as being the first multilateral instrument to be concluded with the prime objective of protecting the environment, and appreciating the significant contribution which that Convention has made in preserving and coastal environment from pollution,

Desiring to achieve the complete elimination of intentional pollution of the marine environment by oil and other harmful substances and the minimization of accidental discharge of such substances,

Considering that this object may best be achieved by establishing rules not limited to oil pollution having a universal purport,

Have agreed as follows:

Article 1

General Obligations under the Convention

1. The Parties to the Convention undertake to give effect to the provisions of the present Convention and those Annexes thereto by which they are bound, in order to prevent the pollution of the marine environment by the discharge of harmful substances or effluents containing such substances in contravention of the Convention.

2. Unless expressly provided otherwise, a reference to the present Convention Constitutes at the same time a reference to its Protocols and to the Annexes.

Article 2

Definitions

For the purposes of the present Convention, unless expressly provided otherwise:

1. "Regulations" means the Regulations contained in the Annexes to the present Convention.

2. "Harmful substance" means any substance which, if introduced into the sea, is liable to create hazards to human health, to harm living resources and marine life, to damage amenities or to interfere with other legitimate uses of the sea and includes any substances subject to control by the present Convention.

3. (a) "Discharge", in relation to harmful substances or effluents containing such substances, means any release howsoever caused from a ship and includes any escape, disposal, spilling, leaking, pumping, emitting or emptying.

(b) "Discharge" does not include:

(i) dumping within the meaning of the Convention on the Prevention of Marine Pollution by Dumping of Wastes and Other Matter done at London on 13 November 1972; or

(ii) release of harmful substances directly arising from the exploration, exploitation and associated off-shore processing of sea-bed mineral resources; or

(iii) release of harmful substances for purpose of legitimate scientific research into pollution abatement or control.

4. "Ship" means a vessel of any type whatsoever operating in the marine environment and includes hydrofoil boats, air-cushion vehicles, submersibles, floating craft and fixed or floating platforms.

5. "Administration" means the Government of the State under whose authority the ship is operating. With respect to a ship entitled to fly a flag of any State, the Administration is the Government of that State. With respect to fixed or floating platforms engaged in exploration and exploitation of the sea-bed and subsoil thereof adjacent to the coast over which the coastal State exercises sovereign rights for the purposes of exploration and exploitation of their natural resources, the Administration is the Government of the coastal State concerned.

6. "Incident" means an event involving the actual or probable discharge into the sea of a harmful substance, or effluents containing such a substance.

7. "Organization" means the Inter-Governmental Maritime Consultative Organization.

Article 3

Application

1. The present Convention shall apply to:

   (a) ships entitled to fly the flag of a Party to the Convention; and

   (b) ships not entitled to fly the flag of a Party but which operate under the authority of a Party.

2. Nothing in the present Article shall be construed as derogating from or extending the sovereign rights of the Parties under International law over the sea-bed and subsoil thereof adjacent to their coasts for the purposes of exploration and exploitation of their natural resources.

3. The present Convention shall not apply to any warship, naval auxiliary or other ship owned or operated by a State and used, for the time being, only on government non-commercial service. However, each party shall ensure by the adoption of appropriate measures not impairing the operations or operational capabilities of such ships owned or operated by it, that such ships act in a manner consistent, so far as is reasonable and practicable, with the present Convention.

Article 4

Violation

1. Any violation of the requirements of the present convention shall be prohibited and sanction shall be established therefore under the law of the Administration of the ship concerned wherever the violation occurs. If the Administration is informed of such a violation and is satisfied that sufficient evidence is available to enable proceedings to be brought in respect of the alleged violation, it shall

cause such proceedings to be taken as soon as possible, in accordance with its law.

2. Any violation of the requirements of the present Convention within the jurisdiction of any party to the Convention shall be prohibited and sanctions shall be established therefore under the law of that Party. Whenever such a violation occurs, that Party shall either:

   (a) cause proceedings to be taken in accordance with its law; or

   (b) furnish to the Administration of the ship such information and evidence as may be in its possession that a violation has occurred.

3. Where information or evidence with respect to any violation of the present Convention by a ship is furnished to the Administration of that ship, the Administration shall promptly inform the Party which has furnished the information or evidence, and the Organization, of the action taken.

4. The penalties specified under the law of a Party pursuant to the present Article shall be adequate in severity to discourage violations of the present Convention and shall be equally severe irrespective of where the violations occur.

Article 5

Certificates and Special Rules on Inspection of Ships

1. Subject to the provisions of paragraph (2) of the present Article a certificate issued under the authority of a Party to the Convention in accordance with the provisions of the Regulations shall be accepted by the other Parties and regarded for all purposes covered by the present Convention as having the same validity as a certificate issued by them.

2. A ship required to hold a certificate in accordance with the provisions of the Regulations is subject while in the ports of off-shore terminals under the jurisdiction of a Party to inspection by officers duly authorized by that Party. Any such inspection shall be limited to verifying that there is on board a valid certificate, unless there are clear grounds for believing that the conditions of the ship or its equipment does not correspond substantially with the particulars of that certificate.

   In that case, or if the ship does not carry a valid certificate, the Party carrying out the inspection shall take such steps as will ensure that the ship shall not sail until it can proceed to sea without presenting an unreasonable threat of harm to the marine environment. That Party may, however, grant such a ship permission to leave the port or off-shore terminal for the purpose of proceeding to the nearest appropriate repair yard available.

3. If a Party denies a foreign ship entry to the ports or off-shore terminals under its jurisdiction or takes any action against such a ship for the reason that the ship does not comply with the provisions of the present Convention, the Party shall immediately inform the consul or diplomatic representative of the Party whose flag the ship is entitled to fly, or if this is not possible, the Administration of the ship concerned.

   Before denying entry or taking such action the Party may request consultation with the Administration of the ship concerned. Information shall also be given to the Administration when a ship does not carry a valid certificate in accordance with the provisions of the Regulations.

Article 6

Detection of Violations and Enforcement of the Convention

1. Parties to the convention shall co-operate in the detection of violations and the enforcement of the provisions of the present Convention, using all appropriate and practicable measures of detection and environmental monitoring, adequate procedures for reporting and accumulation of evidence.

2. A ship to which the present Convention applies may, in any port or off-shore terminal of a Party, be subject to inspection by officers appointed or authorized by that Party for the purpose of verifying whether the ship has discharged any harmful substances in violation of the provisions of the Regulations. If an inspection indicates a violation of the Convention, a report shall be forwarded to the Administration for any appropriate action.

3. Any Party shall furnish to the Administration evidence, if any that the ship has discharged harmful substances or effluents containing such substances in violation of the provisions of the Regulations. If it is practicable to do so, the competent authority of the former Party shall notify the master of the ship of the alleged violation.

4. Upon receiving such evidence, the Administration so informed shall investigate the matter, and, may request the other Party to furnish further or better evidence of the alleged violation, it shall cause such proceedings to be brought in respect of the alleged violation, it shall cause such proceedings to be taken in accordance with its law as soon as possible. The Administration shall promptly inform the Party which has reported the alleged violation, as well as the Organization, of the action taken.

5. A Party may also inspect a ship to which the present Convention applies when it enters the ports of off-shore terminals under its jurisdiction, if a request for an investigation is received from any Party together with sufficient evidence that the ship has discharged harmful substances or effluents containing such substances in any place. The report of such investigation shall be sent to the Party requesting it and to the Administration so that the appropriate action may be taken under the present Convention.

Article 7

Undue Delay to Ships

1. All possible efforts shall be made to avoid a ship being unduly detained or delayed under Articles 4, 5 and 6 of the present Convention.

2. When a ship is unduly detained or delayed under Article 4, 5 and 6 of the present Convention, it shall be entitled to compensation for any loss or damage suffered.

Article 8

Reports on Incidents Involving Harmful Substances

1. A report of an incident shall be made without delay to the fullest extent possible in accordance with the provisions of Protocol I to the present Convention.

2. Each Party to the Convention shall:

   (a) make all arrangements necessary for an appropriate officer or agency to receive and process all reports on incidents; and

   (b) notify the Organization with complete details of such arrangements for circulation to other Parties and Member States of the Organization.

3. Whenever a Party receives a report under the provisions of the present Article, that Party shall relay the report without delay to:

   (a) the Administration of the ship involved, and

   (b) any other State which may be affected

4. Each Party to the Convention undertakes to issue instructions to its maritime inspection vessels and aircraft and to the other appropriate services, to report to its authorities any incident referred to in, Protocol I to the present Convention. That Party shall, if it considers it appropriate, report accordingly to the Organization and to any other Party concerned.

Article 9

Other Treaties and Interpretation

1. Upon its entry into force, the present Convention supersedes the International Convention for the Prevention of Pollution of the Sea by Oil, 1954, as amended, as between Parties to that Convention.

2. Nothing in the present Convention shall prejudice the codification and development of the law of the sea by the United Nations Conference on the Law of the Sea convened pursuant to Resolution 2750C (XXV) of the General Assembly of the United Nations nor the present or future claims and legal views of any State concerning the law of the sea and the nature and extent of coastal and flag State jurisdiction.

3. The term "jurisdiction" in the present Convention shall be construed in the light of international law in force at the time of application or interpretation of the present Convention.

Article 10

Settlement of Disputes

Any dispute between two or more Parties to the Convention concerning the interpretation or application of the present Convention shall, if settlement by negotiation between the Parties involved has not been possible, and if these Parties do not otherwise agree, be submitted upon request of any of them to arbitration as set out in Protocol II to the present Convention.

Article 11

Communication of Information

1. The Parties to the Convention undertake to communicate to the Organization:
   (a) the text of laws, orders, decrees, and regulations and other instruments which have, been promulgated on the various matters within the scope of the present Convention;
   (b) a list of non-governmental agencies which are authorized to act on their behalf in matters relating to the design construction and equipment of ships carrying harmful substances in accordance with the provisions of the Regulations;
   (c) a sufficient number of specimens of their certificates issued under the provisions of the Regulations
   (d) a list of reception facilities including their location, capacity and available, facilities and other characteristics;
   (e) official reports or summaries of official reports in so far as they show the results of the application of the present Convention; and

(f) an annual statistical report, in a form standardized by the Organization, of penalties actually imposed for infringement of the present Convention.

2. The Organization shall notify Parties of the receipt of any communications under the present Article and circulate to all Parties any information communicated to it under sub-paragraphs (1) (b) to (f) of the present Article.

Article 12

Casualties to Ships

1. Each Administration undertakes to conduct an investigation of any casualty occurring to any of its ships subject to the provisions of the Regulations if such casualty has produced a major deleterious effect upon the marine environment.

2. Each Party to the Convention undertakes to supply the Organization with information concerning the findings of such investigation, when it judges that such information may assist in determining what changes in the present Convention might be desirable.

Article 13

Signature, Ratification, Acceptance, Approval and Accession

1. The present Convention shall remain open for signature at the Head quarters of the Organization form 15 January 1974 until 31 December 1974 and shall thereafter remain open for accession. States may become Parties tot the present Convention by:

(a) signature without reservation as to ratification, acceptance or approval; or

(b) signature subject to ratification, acceptance, or approval, followed by ratification, acceptance or approval; or (c) accession.

2. Ratification, acceptance, approval or accession shall be effected by the deposit of an instrument to that effect with the Secretary-General of the Organization.

3. The Secretary-General of the Organization shall inform all states which have signed the present Convention or acceded to it of any signature or of the deposit of any new instrument of ratification, acceptance, approval or accession and the date of its deposit.

Article 14

Optional Annexes

1. A State may at the time of signing, accepting, approving or acceding to the present Convention declare that it does not accept any one or all of Annexes III, IV and V (hereinafter referred to as "Optional Annexes") of the present Convention. Subject to the above, parties to the Convention shall be bound by any Annex in its entirety.

2. A State which has declared that it is not bound by an Optional Annex may at any time accept such Annex by depositing with the Organization and instrument of the kind referred to in Article 13 (2).

3. A State which makes a declaration under paragraph (1) of the present Article in respect of an Optional Annex and which has not subsequently accepted that Annex in accordance with paragraph (2) of the present Article shall not be under any obligation nor entitled to claim any privilege under the present Convention in respect of matters related to such

Annex and all references to Parties in the present Convention shall not include that State in so far as matters related to such Annex are concerned.

4. The Organization shall inform the States which have signed or acceded to the present Convention of any declaration under the present Article as well as the receipt of any instrument deposited in accordance with the provisions of paragraph (2) of the present Article.

Article 15

Entry into Force

1. The present Convention shall enter into force twelve months after the date on which not less than 15 States, the combined merchant fleets of which constitute not less than fifty per cent of the gross tonnage of the world's merchant shipping, have become parties to it in accordance with Article 13.

2. An Optional Annex shall enter into force twelve months after the date on which the conditions stipulated in paragraph (1) of the present Article have been satisfied in relation to that Annex.

3. The Organization shall inform the States which have signed the present Convention or acceded to it of the date on which it enters into force and of the date on which an Optional Annex enters into force in accordance with paragraph (2) of the present Article.

4. For States which have deposited an instrument of ratification, acceptance, approval or accession in respect of the present Convention or any Optional Annex after the requirements for entry into force, thereof have been met but prior to the date of entry into force, the ratification, acceptance, approval or accession shall take effect on the date of entry into

force of the Convention or such Annex or three months after the date of deposit of the instrument whichever is the later date.

5. For States which have deposited an instrument of ratification, acceptance, approval or accession after the date on which the Convention or an Optional Annex entered into force, the Convention or the Optional Annex shall become effective three months after the date of deposit of the instrument.

6. After the date on which all the conditions required under Article 16 to bring an amendment to the present Convention or an Optional Annex into force have been fulfilled, any instrument of ratifications, acceptance, approval or accession deposited shall apply to the Convention or Annex as Amended.

Article 16

Amendments

1. The present Convention may be amended by any of the procedures specified in the following paragraphs.

2. Amendments after consideration by the Organization:

   (a) any amendment proposed by a Party to the Convention shall be submitted to the Organization and circulated by its Secretary-General to all Members of the Organization and all Parties at least six months prior to its consideration;

   (b) any amendment proposed and circulated as above shall be submitted to an appropriate body by the Organization for consideration;

   (c) Parties to the Convention, whether or not Members of the Organization, shall be entitled to participate in the proceedings of the appropriate body;

(d) amendments shall be adopted by a two-thirds majority of only the Parties to the Convention present and voting;

(e) if adopted in accordance with sub-paragraph (d) above, amendments shall be communicated by the Secretary-General of the Organization to all the Parties to the Convention for acceptance;

(f) an amendment shall be deemed to have been accepted in the following circumstances;

(i) an amendment to an Article of the Convention shall be deemed to have been accepted on the date on which it is accepted by two-thirds of the Parties, the combined merchant fleets of which constitute not less than fifty percent of the gross tonnage of the world's merchant fleet;

(ii) an amendment to an Annex to the Convention shall be deemed to have been accepted in accordance with the procedure specified in sub paragraph (f) (iii) unless the appropriate body, at the time of its adoption, determines that the amendment shall be deemed to have been accepted on the date on which it is accepted by two-thirds of the Parties the combined merchant fleets of which constitute not less than fifty per cent of the gross tonnage of the world's merchant fleet. Nevertheless, at any time before the entry into force of an amendment to an Annex to the Convention, a Party may notify the Secretary-General of the Organization that its express approval will be necessary before the amendment enters into force for it. The latter shall bring such notification and the date of its receipt to the notice of Parties;

(iii) an amendment to an Appendix to and Annex to the Convention shall be deemed to have been accepted at the end of a period to be determined by the appropriate body at the time of its adoption, which period shall be not less than ten months, unless within that period an objection is communicated to the Organization by not less than one-third of the Parties or by the Parties the combined merchant fleets of which constitute not less than fifty per cent of the gross tonnage of the world's merchant fleet whichever condition is fulfilled;

(iv) an amendment to Protocol I to the Convention shall be subject to the same procedures as for the amendments to the Annexes to the Convention, as provided for in sub-paragraphs (f) (iii) above;

(v) an amendment to Protocol II to the Convention shall be subject to the same procedures as for the amendments to an Article of the Convention, as provided for in sub-paragraph (f) (i) above;

(g) the amendment shall enter into force under the following conditions:

(i) in the case of an amendment to an Article of the Convention, to Protocol II, or to Protocol I or to an Annex to the Convention not under the, procedure specified in sub-paragraph (f) (iii), the amendment accepted in conformity with the foregoing provisions shall enter into force six months after the date of its acceptance with respect to the Parties which have declared that they have accepted it;

(ii) in the case of an amendment to Protocol I, to an Appendix to an Annex or to an Annex to the Convention under the procedure specified in sub-paragraph (f) (iii), the amendment deemed to have been accepted in accordance with the foregoing conditions shall enter into force six months after its acceptance for all the Parties with the exception of those which, before that date, have made a declaration that they, do not accept it or a declaration under sub-paragraph (f) (ii), that their express approval is necessary.

3. Amendment by a conference:

(a) Upon the request, of a Party, concurred in by at least one-third of the Parties, the Organization shall convene a Conference of Parties to the Convention to consider amendments to the present Convention.

(b) Every amendment adopted by such a Conference by a two-thirds majority of those present and voting of the Parties shall be communicated by the Secretary-General of the Organization to all Contracting Parties for their acceptance.

(c) Unless the Conference decided otherwise, the amendment shall be deemed to have been accepted and to have been entered into force in accordance with the procedures specified for that purpose in paragraph (2) (f) and (g) above.

4. (a) In the case of an amendment to an Optional Annex, a reference in the present Article to a "Party

to the Convention" shall be deemed to mean a reference to a Party bound by that Annex.

(b) Any Party which has declined to accept and amendment to an Annex shall be treated as a non-Party only for the purpose of application of that amendment.

5. The adoption and entry into force of a new Annex shall be subject to the same procedures as for the adoption and entry into force of an amendment to an Article of the Convention.

6. Unless expressly provided otherwise, any amendment to the present Convention made under this Article which relates to the structure of a ship shall apply only to ships for which the building contract is placed, or in the absence of building contract, the keel of which is laid , on or after the date on which the amendment comes into force.

7. Any amendment to a Protocol or to an Annex shall relate to the substances of that Protocol or Annex and shall be consistent with the Articles of the present Convention.

8. The Secretary-General of the organization shall inform all Parties of any amendments which enter into force under the present Article, together with the date on which each such amendment enters into force.

9. Any declaration of acceptance or of objection to an amendment under the present Article shall be notified in writing to the Secretary-General of the Organization. The latter shall bring such notification and the date of its receipt to the notice of the Parties to the Convention.

Article 17

Promotion of Technical Co-operation

The Parties to the Convention shall promote, in consolation with the Organization and other international bodies with assistance and co-ordination by the Executive Director of the United Nations Environment Programme, support for those Parties which request technical assistance for:

(a) the training of scientific and technical personnel;

(b) the supply of necessary equipment and facilities for reception and monitoring;

(c) the facilitation of other measures and arrangements to prevent or mitigate pollution of the marine environment by ships; and

(d) the encouragement of research;

preferably within the countries concerned, so furthering the aims and purposes of the present Convention.

Article 18

Denunciation

1. The present Convention or any optional Annex may be denounced by any Parties to the Convention at any time after the expiry of five years from the date on which the Convention or such Annex enters into force for that Party.

2. Denunciation shall be effected by notification in writing to the Secretary-General of the Organization who shall inform all the Parties of any such notification received and of the date of its receipt as well as the date on which such denunciation takes effect.

3. A denunciation shall take effect twelve months after receipt of the notification of denunciation by the

Secretary-General of the Organization or after the expiry of any other longer period which may be indicated in the notification.

Article 19

Deposit and Registration

1. The present Convention shall be deposited with the Secretary-General of the Organization who shall transmit certified true copies thereof to all States which have signed the present Convention or acceded to it.

2. As soon as the present Convention enters into force, the text shall be transmitted by the Secretary-General of the Organization to the Secretary-General of the United Nations for registration and publication, in accordance with Article 102 of the Charter of the United Nations.

Article 20

Languages

The present Convention is established in a single copy in English, French, Russian and Spanish languages, each text being equally authentic, Official translations in the Arabic, German, Italian and Japanese languages shall be prepared and deposited with the signed original.

In witness whereof the undersigned being duly authorized by their respective Governments for that purpose have signed the present Convention.

Done at London this second day of November, one thousand nine hundred and seventy-three.

4 Helsinki Convention on the Protection of Marine Environment of the Baltic Sea Area (1974)

The States Parties to this Convention

Conscious of the indispensable economic, social and cultural values of the marine environment of the Baltic Sea Area and its living resources for the peoples of the Contracting Parties;

Bearing in mind the exceptional hygrographic and ecological characteristics of the Baltic Sea Area and the sensitivity of its living resources to changes in the environment;

Noting the rapid development of human activities at the Baltic Sea Area, the considerable population living within its catchment area and the highly urbanized and industrialized state of the Contracting Parties as well as their intensive agriculture and forestry;

Noting with deep concern the increasing pollution of the Baltic Sea Area, originating from many sources such as discharges through rivers, estuaries, outfalls and pipelines, dumping and normal operations of vessels as well as through airborne pollutants;

Conscious of the responsibility of the Contracting Parties to protect and enhance the values of the marine environment of the Baltic Sea Area for the benefit of their peoples;

Recognizing that the protection and enhancement of the marine environment of the Baltic Sea Area are tasks that cannot effectively be accomplished by national efforts only but that also close regional cooperation and other appropriate international measures aiming at fulfilling these tasks are urgently needed;

Noting that the relevant recent international conventions even after having entered into force for the respective Contracting Parties do not cover all special requirements to protect and enhance the marine environment of the Baltic Sea Area;

Noting the importance of scientific and technological cooperation in the protection and enhancement of the marine environment of the Baltic Sea Area, particularly between the Contracting Parties;

Desiring to develop further regional cooperation in the Baltic Sea Area; the possibilities and requirements of which were confirmed by the signing of the Convention of Fishing and Conservation of the Living Resources in the Baltic Sea and the Belts, Gdansk 1973;

Conscious of the importance of regional intergovernmental cooperation in the protection of the marine environment of the Baltic Sea Area as an integral part of the peaceful cooperation and mutual understanding between all European States;

Have Agreed as follows;

Article 1

Convention Area

For the purposes of the present Convention "the Baltic

Sea Ara" shall be the Baltic Sea proper with the Gulf of Bothnia, the Gulf of Finland and the entrance to the Baltic Sea bounded by the parallel of the Skaw in the Skagerrak at 57° 44' 8" N. It does not include internal waters of the Contracting Parties.

Article 2

Definitions

For the purposes of the present Convention:

"Pollution" means introduction by man, directly or indirectly, of substances or energy into the marine environment including environment, including estuaries, resulting in such deleterious effects as hazard to human health, harm to living resources and marine life, hinderance to legitimate uses of the sea including fishing, impairment of the quality for use of sea water, and reduction of amenities;

"Land-based pollution" means pollution of the sea caused by discharges from land reaching the sea waterborne, airborne or directly form the coast, including Outfalls froth pipelines;

(a) "Lumping" means:

(i) any deliberate disposal at sea of wastes or other matter from vessels, aircraft, platforms or other man-made structure at sea;

(ii) any deliberate disposal at sea of vessels, aircraft, platforms or other man-made structures at sea;

(b) "Dumping" does not include:

(i) the disposal at sea of wastes or other matter incidentals to, or derived from the normal operations of vessels aircraft, platforms or other man-made structures at sea and their equipment, other than wasted or other matter transported by or to vessels, aircraft, platforms or other man-made structures at sea, operating for the purpose of disposal of such mater or derived from the treatment of such vessels or other matter on such vessels, aircraft, platforms or structures;

(ii) placement of matter for a purpose other than the mere disposal thereof , provided that such placement is not contrary to the aims of the present Convention;

"Vessels and aircraft" means waterborne or airborne craft of any type whatsooever. This expression includes hydrofoil boat.%, air-cushion vehicles, submersibles, floating craft whether self-propelled or not, and fixed or floating platforms;

"Oil" means petroleum in any form including crude oil, fuel oil, sludge, oil refuse and refined products;

"Harmful substance" means any hazardous, noxious, or other substance, which, if introduced into the sea, is liable to cause pollution;

"Incident" means an event involving the actual or probable discharge into the sea of harmful substance, or effluents containing such a substance.

Article 3

Fundamental principles and obligations

1. The Contracting Parties shall individually or jointly take all appropriate legislative, administrative or other relevant measures in order to prevent and abate pollution and to protect and enhance the marine environment of the Baltic Sea Area.
2. The Contracting Parties shall use their best endeavours to ensure that the implementation of the present Convention shall not cause an increase in the pollution of sea areas outside the Baltic Sea Area.

Article 4

Application

The present Convention shall apply to the protection of the marine environment of the Baltic Sea Area which comprises the water-body and the sea-bed including their living resources and other forms of marine life.

Without prejudice to the sovereign rights in regard to their territorial sea, each Contracting Party shall implement the provisions of the present Convention within its territorial sea through its national authorities.

While the provisions of the present Convention do not apply to internal waters, which are under the sovereignity of each contracting Party, the Contracting Parties undertake, without prejudice to their sovereign rights, to ensure that the purposes of the present Convention will be obtained in these waters.

The present Convention shall not apply to any warship, naval auxiliary, military aircraft or other ship and aircraft owned or operated by a State and used, for the time being, only on government non-commercial service.

However, each Contracting Party shall ensure, by the adoption of appropriate measures not impairing the operations or operational capabilities of such ships and aircraft owned or operated by it, that such ships and aircraft act in a manner consistent, so -far as is reasonable and practicable, with the present Convention.

Article 5

Hazardous Substances

The Contracting Parties undertake to counteract the introduction, whether airborne, waterborne or otherwise, into the Baltic Sea Areas hazardous substances as specified in Annex I of the present Convention.

Article 6

Principles and obligations concerning land-based pollution

The Contracting Parties shall take all appropriate measures to control and minimize land-based pollution of the marine environment of the Baltic Sea Area.

In particular, the Contracting Parties shall take all appropriate measures to control and strictly limit pollution by noxious substances and materials in accordance with Annex 11 of the present Convention. To this end they shall, inter alia, as appropriate, cooperate in the development and adoption of specific programmes, guidelines, standards or regulations concerning discharges, environmental quality, and products containing such substances and materials and their use.

The substances and materials listed in Annex II of the present Convention shall not be introduced in to the marine environment of the Baltic Sea Area in significant quantities without a prior special permit, which may be periodically reviewed, by the appropriate national authority.

The appropriate national authority will inform the Commission referred to in Article 12 of the present Convention of the Quantity, quality and way of discharge if it considers that significant quantities of substances and materials listed in Annex II of the present Convention were discharged.

The Contracting Parties shall endeavour to establish and adopt common criteria for issuing permits for discharges.

To control and minimize pollution of the Baltic Sea Area by harmful substances the Contracting Parties shall, in addition to the provision of Article 5 of the present Convention, aim at attaining the goals -And applying the criteria enumerated in Annex III of the -present Convention.

If the discharge from a watercourse, flowing through the territories of two or more Contracting Parties or forming a boundary between them, is liable to cause pollution of the marine environment of the Baltic Sea Area, the Contracting Parties concerned shall in common take appropriate measures in order to prevent and abate such pollution.

The Contracting Parties shall endeavour to use best practicable means in order to minimize the airborne pollution of the Baltic Sea Area by noxious substances.

Article 7

Prevention of pollution from ships

1. In order to protect the Baltic Sea Area from pollution by deliberate, negligent or accidental release of oil, harmful substances other than oil, and by the discharge of sewage and garbage from ships, the Contracting Parties shall take measures as set out in Annex IV of the present Convention
2. The Contracting Parties shall develop and apply uniform requirements for the capacity and location of facilities for the reception of residues of oil, harmful substances other than oil including sewage and garbage, taking into account inter alia the special needs of passenger ships and combination carriers.

Article 8

Pleasure craft

The Contracting Parties shall, in addition to implementing those provisions of the present Convention which can appropriately be applied to pleasure craft, take special measures in order to abate harmful effects on the marine environment of the Baltic Sea Area of pleasure craft activities. The measures shall inter alia deal with adequate reception facilities for wastes from pleasure craft.

Article 9

Prevention of dumping

1. The Contracting Parties shall, subject to Paragraphs 2 and 4 of this Article, prohibit dumping in the Baltic Sea Area.
2. Dumping of dredged spoils shall be subject to a prior special permit by the appropriate national authority in accordance with the provisions of Annex V of the present Convention.
3. Each Contracting Party undertakes to ensure compliance with the provisions of this Article by vessels and aircraft:
   (a) registered in its territory or flying its flag;
   (b) loading, within its territory or territorial sea, matter which is to be dumped; or
   (c) believed to be engaged in dumping within its territorial sea.
4. The provisions of this. Article shall not apply when the safety of human life or of a vessel or aircraft at sea is threatened by the complete destruction or total loss of the vessel or aircraft, or in any case which constitutes a danger to human life, if dumping appears to be the only way of averting the threat and if there is every probability that the damage consequent upon such dumping shall be so conducted as to minimize the likelihood of damage to human or marine life.
5. Dumping made under the provisions of Paragraph 4 of this article shall be reported and dealt with in accordance with Annex VI of the present Convention and shall also be reported forthwith to the Commission referred to in Article 12 of the present Convention in accordance with the provisions of Regulation 4 of Annex V of the present Convention.

6. In case of dumping suspected to be in contravention of the provisions of this Article the Contracting Parties shall co-operate in investigating the matter in accordance with Regulation 2 of Annex IV of the present Convention.

Article 10

Exploration and exploitation of the sea-bed and its subsoil

Each Contracting Party shall take all appropriate measures in order to prevent pollution of the marine environment of the Baltic Sea Area resulting from exploration or exploitation of its part of the sea-bed and its subsoil or form any associated activities thereon. It shall ensure that adequate equipment is at hand to start an immediate abatement of pollution in that area.

Article 11

Co-operation in combatting marine pollution

The Contacting Parties shall take measures and cooperate as set out in Annex VI of the present Convention in order to eliminate or minimize pollution of the Baltic Sea Area by oil or other harmful substances.

Article 12

Institutional and organizational framework

1. The Baltic Marine Environment Protection Commission, hereinafter referred to as "the Commission", is hereby established for the purposes of the present Convention.
2. The chairmanship of the Commission shall be given to each Contracting Party in turn in alphabetical order to the names of the States in the English language.

The Chairman shall serve for a period of two years, aid cannot during the period of his chairmanship serve as representative of his country.

Should the chairmanship fall vacant, the Contracting Party chairing *the Commission* shall nominate a successor to remain in office until the term of chairmanship of that Contracting Party expires.

3. Meetings of the Commission shall be hold at least once a year upon convocation by the Chairman. Upon request of a Contrating Party, provided it is endorsed by another Contracting Party, the Chairman shall, as soon as possible, summon an extraordinary meeting at such time and place as the Chairman determines, however, not later than ninety days from the date of the submission of the request.
4. The first meeting of the Commission shall be called by the Depositary Government and shall take place within a period of ninety days from the date following the entry into force of the present Convention.
5. Each Contracting Party shall have not one vote in the Commission. Unless otherwise provided under the present Convention, the Commission shall take its decisions unanimously.

Article 13

The duties of the Commission shall be;

(a) To keep the implementation of the present Convention under continuous observation;

(b) To make recommendations on measures relating to the purposes of the present Convention,

(c) To keep under review the contents of the present Convention including its Annexes and to recommend to the Contracting Parties such amendments to the present Convention including its Annexes as may be required including changes in the lists of substances and materials as well as the adoption of new Annexes;

(d) To define pollution control criteria, objectives for the reduction of pollution, and objectives concerning measures, particularly according to Annex III of the present Convention;

(e) To promote inclose cooperation with appropriate governmental bodies, taking into consideration subparagraph; (f) of this Article, additional measures to protect the marine environment of the Baltic Sea Area and for this purpose:

(i) to receive, process, summarize and disseminate from available sources relevant scientific, technological and statistical information; and

(ii) to promote scientific and technological research;

(f) To seek, when appropriate, the services of competent regional and other international organizations to collaborate in scientific and technological research as well as other relevant activities pertinent to the objectives of the present Convention;

(g) To assume such other functions as may be appropriate under the terms of the present Convention.

Article 14

Administrative provisions for the Commission

1. The working language of the Commission shall be English

2. The Commission shall adopt its Rules of Procedure.

3. The office of the Commission, hereafter referred to as the "Seecretariat", shall be in Helsinki.

4. The Commission shall appoint an Executive Secretary and make provisions for the appointment of such other personnel as may be necessary, and determine the duties, terms and conditions of the Executive Secretary.

5. The Executive Secretary shall be the chief administrative official of the Commission and shall perform the function that are necessary for the administration of the present Convention, the work of the Commission and other tasks entrusted to the Executive Secretary by the Commission and its Rules of Procedure.

Article 15

Financial provisions for the Commission

1. The Commission shall adopt its Financial rules.
2. The Commission shall adopt an annual or biennial budget of proposed expenditures and budget estimates for the fiscal period following thereafter.
3. The total amount of the budget, including any supplementary budget adopted by the Commission, shall be contributed by the Contracting Parties in equal parts, unless the Commission unanimously decides otherwise.
4. Each Contracting Party shall pay the expenses related to the participation in the Commission of its representatives, experts and advisers.

Article 16

Scientific and technological cooperation

1. The Contracting Parties undertake directly, or when appropriate through competent regional or other international organizations, to cooperate a the fields of science, technology and other research, and to exchange date as well as other scientific information for the present Convention.
2. Without prejudice to Paragraphs 1,2 and 3 of Article 4 of the present Convention the Contracting Parties undertake directly, or when appropriate, through competent regional or other international organizations to promote studies, undertake, support

or contribute to programmes aimed at developing ways and means for the assessment of the nature and extent of pollution, pathways, exposures~ risks and remedies in the Baltic Sea Area, and particularly to develop alternative methods of treatment, disposal and elimination of such matter and substances that are likely to cause pollution of the marine environment of the Baltic Sea Area.

3. The Contracting Parties undertake directly, or when appropriate through competent regional or other international organizations', and on the basis of the information and data acquired pursuant to Paragraphs 1 and 2 of this Article, to cooperate in developing intercomparable observation methods, in performing baseline studies and in establishing complementary or joint programmes for monitoring.

4. The organization and scope of work connected with the implementation of tasks referred to in the preceding Paragraphs should primarily be outlined by the Commission.

Article 17

Responsibility for damage

The Contracting Parties undertake, as soon as possible, jointly to develop and accept rules concerning responsibility for damage resulting from acts of omissions in contravention of the present Convention, including, inter alia, limits of responsibility, criteria and procedures for the determination of liability and available remedies.

Article 18

Settlement of disputes

1. In case of a dispute between Contracting Parties as to the interpretation or application of the present Convention, they should seek a solution by

negotiation. If the Parties concerned cannot reach agreement they should seek the good offices of or jointly request the mediation by a third Contracting Party, a qualified international organization or a qualified person.

2. If the Parties concerned have not been able to resolve their dispute through negotiation or have been unable to agree on measures as described above, such disputes shall be, upon common-agreement, submitted to an ad hoc arbitration tribunal, to a permanent arbitration tribunal, or to the International Court of Justice.

Article 19

Safeguard of certain freedoms

Nothing in the present Convention shall be construed as infringing upon the freedom of navigation, fishing, marine scientific research and other legitimate uses of the high seas, as well as upon the right of innocent passage through the territorial sea.

Article 20

Status of Annexes

The Annexes attached to the present Convention form an integral part of the Convention.

Article 21

Relation to other Conventions

The provisions of the present Convention shall be without prejudice to the rights and obligations of the Contracting parties under treaties concluded previously as well as under treaties which may be concluded in the future, furthering and developing the general principles of the Law of the Sea that the present. Convention is based upon and in particular provisions concerning the prevention of pollution of the marine environment.

Article 22

Revision of the Convention

A conference for the purpose of a general revision of the present Convention may be convened with the consent of the Contracting Parties or at the request of the Commission.

Article 23

Amendments to the Articles of the Convention

1. Each Contracting Party may propose amendments to the Articles of the present Convention. Any such proposed amendment shall be submitted to the Depositary Government and communicated by it to all Contracting Parties, which shall inform the Depositary Government of either their acceptance or rejection of the amendment as soon as possible after the receipt of the communication.

The amendment shall enter into force ninety days after the Depositary Government has received notifications of acceptance of that amendment from all Contracting Parties.

2. With the consent of the Contracting Parties or at the request of the Commission a conference may be convened for the purpose of amending the present Convention.

Article 24

Amendements to the Annexes and the adoption of Annexes

1. Any amendment to the Annexes proposed by a Contracting Party shall be communicated to the other Contracting Parties by the Depositary Government and considered in the Commission. If adopted by the Commission, the amendment shall be communicated to the Contracting Parties and recommended for acceptance.

2. Such amendment shall be deemed to have been accepted at the end of a period determined by the Commission unless within that period anyone of the Contracting Parties has objected to the amedment. The accepted amendment shall enter into force on a date determined by the Commission.

The period determined by the Commission shall be prolonged for an additional period of six months and the date of entry in to force of the amendment posfponed accordingly, if, in exceptional cases, any Contracting Party before the expiring of the period determined by the Commission informs the Depositary Government, that, although it intends to accept the proposal, the constitutional requirements for such an acceptance are not yet fulfilled in its State.

3. An Annex to the present Convention may be adopted in accordance with the provisions of this Article.
4. The Depositary Government shall inform all Contracting Parties of any amendments or adoption of a new Annex which enter into force under this Article and of the date on which such amendment or new Annex enters into force.
5. Any objection under this Article shall be made by notification in writing to the Depositary Government which shall notify all Contracting Parties and the Executive Secretary of any such notification and the date of its receipt.

***Article* 25**

Reservations

1. The provisions of the present Convention shall not be subject to reservations.
2. The provision of Paragraph 1 of this Article does not prevent a Contracting Party from suspending for a

period not exceeding one year the application of an Annex of the present Convention or part thereof or an amedment thereto after the Annex in question or the amendment thereto has entered into force.

3. If after the entry into force of the present Convention a Contracting Party invokes the provisions of Paragraph 2 of this Article it shall inform the other Contracting Parties, at the time of the adoption by the Commission of an amendment to an Annex or a new Annex, of those provisions which will be suspended in accordance with Paragraph 2 of this Article.

Article 26

Signature, ratification, approval, and accession

1. The present Convention shall be open for signature in Helsinki on 22 March 1974 by the Baltic Sea States participating in the Diplomatic Conference on the Protection of the Marine Environment of the Baltic Sea Area, held in Helsinki from 18 to 22 March 1974. The present Convention shall be open for accession to any other Convention, provided that this State is invited by all the Contracting Parties.
2. The present Convention shall be subject to ratification or approval by the States which have signed it.
3. The instrument of ratification, approval, or accession shall be deposited with the Government of Finland, which will perform the duties of the Depositary Government.

Article 27

Entry into force

The present Convention shall enter into force two months after the Depositary of the seventh instrument of ratification on approval.

Article 28

Withdrawal

1. At any time after the expiry of five years from the date of entry into force of the present Convention any Contracting Party may, by giving written notification to the Depositary Government was notified of the withdrawal.
2. In case of notification of withdrawal by a Contracting Party the Depositary Government shall convene a meeting of the Contracting Parties for the purpose of considering the effect of the withdrawal.

Article 29

Language

The present Convention has been drawn up in a single copy in the English language. Official translations into the Danish, Finnish, German, Polish, Russian, and Swedish languages shall be prepared and deposited with the signed original.

In witness whereof the undersigned Plenipotentiaries, being duly authorized thereto, have signed the present Convention.

Done at Helsinki, this twenty-second day of March one thousand nine hundred and seventy-four.

Annex-I

Hazardous Substances

The protection of the Baltic Sea Area from pollution by the substances listed below can involve the use of appropriate technical means, prohibitions and regulations of the transport, trade, handling, application, and final deposition of products containing such substances.

1. DDT (1,1,1-trichloro-2,2-bis-(chlorophenyl)-ethane) and its derivatives DDE and DDD.

2. PCBs(polychlorinated biphenyls).
3. PCT's (polycholorinated terphenyls).

Annex II

Noxious Substances and Materials

The following substances and materials are listed for the purposes of Article 6 of the present Convention.

The list is valid for substances and materials introduced as waterborne into the marine environment. The Contracting Parties shall also endeavour to use best practicable means to prevent harmful substances and materials form being introduced as airborne into the Baltic Sea Area.

For urgent consideration

1. Mercury, cadmium, and their compounds.
2. Antimony, arsenic, beryllium, chromium, copper, lead, molybdenum, nickel, selenium, tin, vanadium, zinc and their compounds, as well as elemental phosphorus
3. Phenols and their derivatives.
4. Phthalic acid and its derivatives.
5. Cyanides.
6. Persistent halogenated hydrocarbons.
7. Polycyclic aromatic hydrocarbons and their derivatives.
8. Persistent toxic organosilicic compounds.
9. Persistent pesticides, including organophosphoric and organostannic pesticides, herbicides, slimicides and chemicals used for the preservation of wood, timber, wood pulp, cellulose, paper, hides and textiles, not covered by the provisions of Annex 1 of the present Convention.
10. Radioactive materials.

11. Acids, alkalis, and surface active agents in high concentration or big quantities.
12. Oil and wastes of petrochemical and other industries containing lipid-soluble substances.
13. Substances having adverse effects on the taste and/or smell of products for human consumption from the sea, or effects on taste, smell, colour, transparency or other characteristics of the water seriously reducing its amenity values.
14. Materials and substances which may float, remain in suspension or sink, and which may seriously interfere with any legitimate use of the sea.
15. Lignin substances contained in industrial waste waters.
16. The chelators EDTA (ethylenedinitrilotetraacetic acid or ethylenediaminetetraacetic acid) and DTPA (diethylenetri-aminopentaacetic acid).

Annex III

Goals, Criteria and Measures concerning the Prevention of Land-based Pollution

In accordance with the provisions of Article 6 of the present Convention the Contracting Parties shall, endeavour to attain the goals and apply the criteria and measures enumerated in this Annex in order to control and minimize land-based pollution of the marine environment of the Baltic Sea Area.

1. Municipal sewage shall be treated in an appropriate way so that the amount of organic matter does not cause harmful changes in the oxygen content of the Baltic Sea Area.
2. Municipal sewage shall also be treated in an appropriate way to ensure that the hygienic quality, and in particular epidemiological and toxicological safety, of the receiving sea area is maintained at a

level which does not cause harm to human health, and in a way that under the given composition of the sewage no significant amount of such harmful substances as are listed in Annexes I and II of the present Convention is formed.

3. The polluting load of industrial wastes shall be minimized in an appropriate way in order to reduce the amount of harmful substances, organic matter and nutrients.
4. The means referred to in Paragraph 3 of this Annex shall in particular include minimization of production of wastes by processing techniques, recirculation and re-use of processing water, developing of water economy and improvement of qualifications for water treatment. In the treatment of waste water mechanical, chemical biological and other measures, according to the quality of the recipient water, shall be applied.
5. The discharge of cooling water from nuclear power plants or other kinds of industries using large amounts of water shall be effected in a way which minimizes the pollution of the marine environment of the Baltic Sea Area.
6. The Commission will control criteria, objectives for reduction of pollution and objectives concerning measures, including processing techniques and waste treatment, to reduce pollution of the Baltic Sea Area.

Annex IV

Prevention of Pollution from Ships

Regulation 1

The contracting Parties shall, in matters concerning the protection of the Baltic Sea Area from pollution by ships, cooperate.

within the International Martime Organization, in particular in promoting the development of international rules,

in the effective and harmonized implementation of rules adopted by the International Maritime Organization.

Regulation 2

The Contracting Parties shall, without prejudice to Paragraph 4 of Article 4 of the present Convention, as apropriate assist each other investigating violations of existing legislation on anti-pollution measures, which have occurred or are suspected to have occurred within the Baltic Sea Area. This assistance may include but is not limited to inspection by the competent authorities of oil record books, cargo record books, log books and engine log books and taking oil samples for analytical identification purposes.

Regulation 3

Definitions

For the purpose of this Annex:

1. "Ship" means a vessel of any type whatsoever operating in the marine environment and includes hydrofoil boats, air-cushion vehicles, submersibles, floating craft and fixed or floating platforms.
2. "Administration" means the Government of the State under whose authority the ship is operating. With respect to a ship entitled to fluy a flag of any State, the Administration is the Government of that State. With respect to fixed or floating platforms engaged in exploration and exploitation of the sea-bed and subsoil thereof adjacent to the coast over which the costal State exercises sovereign rights for the purposes of exploration and exploitation of

their natural resources, the Administration is the Government of the Coastal State concerned.

3. (a) "Discharge", in relation to harmful substances or effluents containing such substances, means any release howsoever caused form a ship and includes any escape, disposal, spilling, leaking, pumping, emitting, or emptying; (b) "Discharge" doesnot include:

   (i) dumping within the meaning of the Convention on the Prevention of Marine Pollution by Dumping of Wastes and other Matter done at London on 29 December 1972; or

   (ii) release of harmful substances directly arising from the exploration, exploitation and associated offshore processing of sea-bed mineral resources; or

   (iii) release of harmful substances for purposes of legitimate scientific research into pollution abatement or control .

4. "Nearest land" The term "from the nearest land" means from the baseline from which the territorial sea of the territory in question is established in accordance with international law.

5. The term 'Jurisdiction" shall be interpreted in accordance with international law in force at the time of application or interpretation of this Annex.

6. The term "MARPOL 73/78" means the International Convention for the Preventions of Pollution from Ships, 1973, as modified by the Protocol of 1978 relating thereto.

Regulation 4

Oil

The Contracting Parties, also being parties to MARPOL

73/78, apply in conformity with that agreement the provisions of Annex I to MARPOL 73/78 for the prevention of pollution by oil.

Definitions

For the purposes of this Regulation:

1. "Oil" means petroleum in any form including crude oil, fuel oil, sludge, oil refuse and refined products (other than petrochemicals which are subject to the provision of Regulation 5 of this Annex) and, without limiting the generality of the foregoing, includes the substances listed in Appendix I to this Annex.
2. "Oily mixture" means a mixture with any oil content.
3. "Oil fuel" means any oil used as fuel in connection with the propulsion and auxiliary machinery of the ship in which such oil is carried.
4. "Oil tanker" means a ship constructed or adapted primarily to carry oil in bulk in its cargo spaces and includes combination carriers and any "chemical tanker" as defined in Regulation 5 of this Annex when it is carrying a cargo or part cargo of oil in bulk.
5. "Combination carrier" means a ship designed to carry either oil or solid cargos in bulk.
6. "Clean ballast" means the ballast in a tank which since oil was last carried therein has been so cleaned that effluent therefrom if it were discharged from a ship which is stationary into clean calm water on a clear day would not produce visible traces of oil on the surface of the water or on adjoining shore lines or cause a sludge or emulsion to be deposited beneath the surface of the water or upon adjoining shore lines. If the blast is discharged through an oil

discharge monitoring and control system approved by the Administration, evidence based on such a system to the effect that the oil content of the effluent did not exceed 15 parts per million shall be determinative that the ballast was clean, notwithstanding the presence of visible traces.

7. "Segregated Ballast" means the ballast water introduced into a tank which is completely separated from the cargo oil and oil fuel system and which is permanently allocated to the carriage of ballast or to the carriage of ballast or cargos other than oil or noxious substances as variously defined in the Regulations of this Annex.

Control Discharge of Oil

1. (a) Subject to the provision of Paragraph C of this Regulation, any discharge into the sea of oil or aoily mixtures from any oil tanker and any ship of 400 tons gross tonnage and above other than an oil tanker shall be prohibited, while in the Baltic Sea Area;

   (b) such ships. while in the Baltic Sea Area shall retain on board all oil drainage and sludge, dirty ballast and tank washing waters and discharge them only to reception facilities.

2. (a) Subject to the provision of Paragraph C of this Regulation, any discharge into the sea of oil or oily mixtures from a ship of less than 400 tons gross tonnage, other than an oil tanker, shall be prohibited while in the Baltic Sea Area, except when the oil content of the effluent without dilution does not exceed 15 parts per million or alternatively when all of the following conditions are satisfied:

   (i) the ship is proceeding en route;

(ii) the oil content of the effluent is less than 100 parts per million; and

(iii) the discharge is made as far as practicable from the land, but in no case less than 12 nautical miles from the nearest land;

3. (a) The provisions of Sub-Paragraphs 1 and 2 of this Paragraph shall not apply to the discharge of clean or segregated ballast.

(b) The provisions of Sub-Paragraph 1 of this Paragraph shall not apply to the discharge of processed bilge water from machinery, spaces, provided that all of the following conditions are satisfied:

(i) the bilge water does not originate from cargo pump;

(ii) the bilge water is not mixed with oil cargo residues;

(iii) the ship is proceeding en route;

(iv) the oil content of the effluent without dilution does not exceed 15 parts per million;

(v) the ship has in operation an oily-water separating system and an effective filtering system, or an equivalent equipment, approved by the Administration;

(vi) the system or equipment is such that it will produce an effluent, the oil content of which does not exceed 15 parts per million, and is provided with alarm arrangements to indicate when this level cannot maintained, as well as a stopping be device which will ensure that the discharge is automatically stopped when the oil content of the effluent exceeds 15 parts per million.

4. (a) No discharge into the sea shall contain chemicals or other substances in quantities or concentration which are hazardous to the marine environment of chemicals or other substances introduced for the purpose of circumventing the conditions of discharge specified in this Regulation.

   (b) The oil residues which cannot be discharged into the sea in compliance with Sub-Paragraphs 2 or 3, (b) of .this Paragraph shall be retained on board or discharged to reception facilities.

5. Whenever visible traces of oil are observed on or below the surface of the water in the immediate vicinity of a ship or its wake, the Contracting Parties should, to the extent they are reasonably able to do so, promptly investigate the facts bearing on the issue of whether there has been a violation of the provisions of this Regulation.

The Investigation should include, in particular, the wind and sea conditions, the track and speed of the ship, other possible sources of the visible traces in the vicinity, and any relevant oil discharge records.

Exceptions

Paragraph B of this Regulation shall not apply to:

(a) the discharge into the sea of opil or oily mixtures necessary for the purpose of securing the safety of a ship or saving life at sea; or

(b) the discharge into the sea of oil or oily mixtures resulting from damage to a ship or its equipment:

   (i) provided that all reasonable precautions have been taken after the occurrence of the damage or discovery of the discharge for the purpose of preventing or minimizing the discharge; and

   (ii) except if the owner or the Master acted either with intent to cause damage, or recklessly and with knowledge that damage would probably result; or

(c) the discharge into the sea of substances containing oil, approved by the Administration, when being used for the purposes of combating specific pollution incidents in order to minimize the damage from pollution. Any such discharge shall be subject to the approval of any Contracting Party in whose jurisdiction it is contemplated discharge will occur.

Special Requirements for Drilling Rigs and other Platforms

Fixed and floating drilling rigs when engaged in the exploration, exploitation and associated offshore processing of sea-bed mineral resources and other platforms shall comply with the requirements of this Regulation applicable to ships of 400 tons gross tonnage and above other than oil tankers, except that:

(a) they shall keep a record of all operations involving oil or oily mixture discharges, in a form approved by the Administration; and

(b) subject to the provision of Paragraph C of this Regulation, the discharge into the sea of oil or oily mixture shall be prohibited except when the oil content of the discharge without dilution does not exceed 15 parts per million.

Reception Facilities of Baltic Sea Area

The Contracting Parties undertake to ensure that no later than 1 January 1977 all oil loading terminals an repair ports of the Baltic Sea Area are -provided wit facilities adequate for the reception and treatment of all the dirty ballast and tank washing waters from oil tankers In addition all ports of the area shall be provided with adequate reception facilities for other residues and oily mixtures from all ships. Such facilities shall have adequate capacity to meet the needs of the ships using them without causing undue delay.

Regulation 5

Noxious liquid substances

The Contracting Parties, also being parties to ARPOL 73/ 78., apply in conformity with that agreement the provisions of Annex 11 to ARPOL 73/78 for the prevention of pollution by noxiouss liquid substances carried in bulk. A Definitions

1. "Chemical tanker" means a ship constructed or adapted primarily to carry a cargo of noxious liquid substances in bulk and includes an "oil tanker" as defined in Regulation 4 of this Annex when carrying a cargo or part cargo of noxious liquid substances in bulk.
2. "Clean Ballast" means ballast carried in a tank which, since it was last used to carry a cargo containing a substance in Category A,B,C, or D has been thoroughly cleaned and the residues resulting therefrom have been discharged and the tank emptied in accordance with the appropriate requirements of this Regulation.
3. "Segregated ballast" means ballast water introduced into a tank permanently allocated to the carriage of ballast or to the carriage of ballast or cargos other than oil or noxious liquid substances as variously defined in the Regulations of this Annex, ' and which is completely separated form the cargo and oil fuel system.
4. "Liquid substances" are having a vapour pressure not exceeding 28 kp/cm at a temperature of 37.8 C
5. "Noxious liquid substances" means any substances designated in Appendix III to this Annex or provisionally assessed under the provisions of Sub-Paragraph 4 of paragraph B of this Regulation as falling into Category A,B,C, or D.

Categorization and Listing of Noxious Liquid Substances

1. For the purposes of this Regulations noxious liquid substances shall be divided into four categories as follows:
   (a) Category A—noxiousa liquid substances which if discharged into the sea from tank cleaning or deballasting operations would present a major hazard to either marine resources or human health or cause serious harm to amenities or other legitimate uses of the sea and therefore justify the application of stringent anti-pollution measures;
   (b) Category B—noxious liquid substances which if discharged into the sea from tank c leaning or deballasiting operations would present a hazard to either marine resources or human health or cause harm to amenities or other legitimate uses of the sea and therefore justify the application of special anti-pollution measures;
   (c) Category C—noxious liquid substances which if discharged into the sea from tank cleaning or deballasting operations would present a minor hazard to either marine resources or human health or cause, minor harm to amenities or other legitimate uses of the sea and therefore require special operational conditions;
   (d) Category D—noxious liquid substances which if discharged into the sea from tank cleaning or deballasing operations would present a recognizable hazard to either marine resources or human health or cause minimal harm to amenities or other legitimative uses of the sea an therefore require some attention in operational conditions.

2. Guidelines for use in the categorization of noxious liquid substances are given in Appendix II to this Annex.

3. The list of noxious liquid substances carried in bulk and presently categorized which are subject to the provisions of this Regulation is set out in Appendix II to this Annex.

4. Where it is proposed to carry a liquid substance in bulk which has not been categorized unser Sub-Paragraph 1 of this Paragraph or evaluated as referred to in Sub-Paragraph 1 of Paragraph C of this Regulation the Contracting Parties involved in the proposed operation shall establish and agree on a provisional assessment for the proposed operation on the basis of the guidelines referred to in Sub-Paragraph 2 of this Paragraph. Until full agreement between the Governments involved has been reached, the substances shall be carried under the most severe conditions proposed.

C Other Liquid Substances

The substances listed in Appendix IV to this Annex have been evaluated and found to fall outside the Categories A,B,C, and D as defined in Sub-Paragraph 1 of Paragraph B of this Regulation because they are presently considered to present no harm to human health marine resources, amenities or other legitimate uses of the sea, when discharged into the sea from tank cleaning or deballasting operations. The discharge of bilge or ballast or other residues of mixtures containing oily substances listed in Appendix IV to this shall not be subject to any requirement of this' Regulation.

The discharge into the sea of clean ballast or segregated ballast shall not be subject to any requirement of this Regulation

Discharge of Noxious Liquid Substances

Subject to the provisions of Paragraph E of this Regulation:

1. The discharge into the sea of substances in Category A as defined in Sub-Paragraph 1 (a) of a Paragraph B of this Regulation, or of those provisionally assessed as such or ballast water, tank washings, or other residues or mixtures containing such substances shall be prohibited. If tanks containing such substances or mixtures are to be washed, the resulting residues shall be discharged to a reception facility which the Contracting Parties shall provide in accordance with Paragraph H of this Regulation, until the concentration of the substance in the effluent to such facility is at or below the residual concentration prescribed for that substance in column IV of Appendix III to this Annex and until the tank is empty. Provided that the residue then remaining in the tank is subsequently diluted by the addition of a volume of water of not less than 5 per cent of the total volume of the tank, it may be discharged into the sea when all the following conditions are also satisfied:
   (a) the ship is proceeding en route at a speed of at least 7 knots in the case of self-propelled ships or at least 4 knots in the case of ships which are not self-propelled;
   (b) the discharge is made below the waterline, taking into account the location of the seawater intakes; and
   (c) the discharge is made at a distance of not less than 12 nautical miles from the nearest land and in a depth of water not less than 25 metres.

2. The discharge into the sea of substances in Category B as defined in Sub- Paragraph I (b) of Paragraph B of this Regulation or of those provisionally assessed as such or ballast water, tank washings, or other residues or mixtures containing such substances shall be prohibited except when all the following conditions are satisfied:

   (a) the tank has been washed after unloading with a volume of water of not less than 0.5 percent of total volume of the tank, and the resulting residues have been discharged to a reception facility until the tank is empty;

   (b) the ship is proceeding en route at a speed of at least 7 knots in the case of self-propelled ships or at least 4 knots in the a case of self-propelled ships or at least 4 knots in the case of ships which are not self-propelled.

   (c) the procedures and arrangements for discharge and washings are approved by the Administration and shall ensure that the concentration and rate of discharge of the effluent is such that the concentration of the substance in the wake astern of the ship does not exceed part per million.

   (d) the discharge is made below the waterline; taking into account the location of the seawater intakes; and

   (e) the discharge is made at distance of not less than 12 nautical miles form the nearest land and in depth of water of not less than 25 meteres.

3. The discharge into the sea of substances in Category C as defined in Sub-Paragraph I (c) of Paragraph B of this Regulation or of those provisionally assessed as such or ballast water, tank washings, or other

residues or mixtures containing such substances shall be prohibited except when all the following conditions are satisfied:

(a) the ship is proceeding en route at a speed of at least 7 knots in the case of self-propelled ships or at least 4 knots in the case of ships, which are not self-propelled;

(b) the procedures and arrangements for discharge are approved by the Administration and shall ensure that the concentration and rate of discharge of the effluent is such that the concentration of the substance in the wake astern of the ship does not exceed I part er million

(c) the maximum quantity of cargo discharged from each tank and its associated piping system does not exceed the maximum quantity approved in accordance with the procedures referred to in Sub-Paragraph 3 (b) of this Paragraph which shall not case exceed the greater of 1 cubic metere of 1/ 3,0000 of the tank capacity in cubic metres;

(d) the discharge is made below the waterline, taking into account the location of the seawater intakes; and

(e) the discharge is made at a distance of not less than 12 nautical miles from the nearest land and in a depth of water of not less than 25 metres.

The discharge into the sea of substances in Category D as defined in Sub-Paragraph 1 (d) of Paragraph B of this Regulation, or of those provisionally assessed as such or ballast water, tank washings, or other residues of mixtures containing such substances shall be prohibited except when all the following conditions are satisfied:

(a) the ship is proceeding en route at a speed of at least 7 knots in the case of self-propelled ships or at least 4 knots in the case of ships which are not self-propelled;

(b) such mixtures are of a concentration not greater than one part of the substance in ten parts of water; and

(c) the discharge is made at a distance of not less than 12 nautical miles from the nearest land.

Ventilation procedures approved by the Administration, may be used to remove cargo residues from a tank. If subsequent washing of the tank is necessary, the discharge into the sea of the resulting tank washings shall be made in accordance with Sub-Paragraph 1, 2, 3 or 4 of this Paragraph, whichever is applicable.

The discharge into the sea of substances which have not been categorized, provisionally assessed, or evaluated as referred to in Sub-Paragraph 1 of Paragraph C of this Regulation, or of ballast water, tank washings, or other residues of mixtures containing such substances shall be prohibited.

Exceptions

Paragraph D of this Regulation shall not apply to:

(a) the discharge into the sea of noxious liquid substances or mixtures containing such substances necessary for the purpose of securing the safety of a ship or saving life at sea; or

(b) the discharge into the sea of noxious liquid substances or mixtures containing such substances resulting from damage to, a ship or its equipment:

(i) provided that all reasonable precautions have been taken after the occurrence of the damage or discovery of the discharge for the purpose of preventing or minimizing the discharge; and

(ii) except if the owner or the Master acted either with intent to cause damage, or recklessly and with knowledge that damage would probably result; or

(c) the discharge into the sea of noxious liquid substances or mixtures containing such substances, approved by the Administration, when being used for the purpose of combatting specific pollution incidents in order to minimize the damage from pollution. Any such discharge shall be subject to the approval of any Contracting Party in whose jurisdiction it is contemplated the discharge will occur.

Measures of Control

1. The Contracting parties shall appoint or authorize surveyors for the purpose of implementing this Paragraph.

Category A Substances

2. (a) If a tank is partially unloaded or unloaded but not cleaned, an appropriate entry shall be made in the Cargo Record Book;

   (b) until that tank is cleaned every subsequent pumping or transfer operation carried out connection whit that tank shall also be entered the Cargo Record Book.

3. If the tank is to be washed:

   (a) the effluent from the tank washing operation shall be discharged from the ship to reception facility at least until the concentration of the substance in the discharge, as indicated by analyses c samples of the effluent taken the surveyor, ha fallen to the residual concentration specified that substance in Appendix III to this Annex When the required residual concentration has bee achieved remaining tank washings shall continue to be discharged to the reception facility until the tank is empty. Appropriate entries of these operations shall be made in the Cargo Record Book and certified by the surveyor; and

(b) after diluting the residue then remaining in the tank, with at least. 5 per cent of the tank capacity of water, this mixture may be discharged into the sea in accordance with the provisions of Sub-Paragraphs 1 a), b) and c) of Paragraph D of this Regulation. Appropriate entries of these operations shall be made in the Cargo Record Book.

4. Where the Government of the receiving Party is satisfied that it is impracticable ~to measure the. concentration of the substance in the effluent without causing undue delay to the ship, that Party may accept an alternative procedure as being equivalent to Sub-Paragraph 3 a) of this Paragraph provided that:

(a) a precleaning procedure for that tank and that substance is approved by -the Administration and that Party is satisfied that such procedure will fulfil the requirements of Sub-Paragraph 1, of Paragraph D of this Regulation with respect to the attainment of the prescribed residual concentrations;

(b) a surveyor duly authorized by that Party shall certify in the Cargo Record Book that:

(i) the tank, its pump and piping systems have been emptied and that the quantity of cargo remaining in the tank is at or below the quantity on which the approved precleaning procedure referred to in Sub-Paragraph (ii) of this Sub-Paragraph has been based;

(ii) precleaing has been carried out in accordance with the precleaning procedure approved by the Administration for that tank and that substance; and

(iii) the tank washings resulting from such precleaning have been discharged to a reception facility and the tank is empty;

(c) the discharge into the sea of any remaining residues shall be, in accordance with the provisions of Sub-Paragraph 3 b) of this Paragraph and an appropriate entry is made in the Cargo Record Book.

Category B Substances

5. Subject to such surveillance and approval by the authorized or appointed surveyor as may be deemed necessary by the Contracting Party, the Master of a ship shall, with respect to a Category B Substance, ensure compliance with the following:

(a) if a tank is partially unloaded or unloaded but not cleaned appropriate entry shall be made in the Cargo Record Book;

(b) until that tank is cleaned every subsequent pumping or transfer operation carried out in connection with that tank shall also be entered in the Cargo Record Book;

(c) if the tank is to be washed, the effluent from the tank washing operation, which shall contain a volume of water not less than 0.5 per cent of the ,total volume of the tank, its pump and piping system are empty. An appropriate entry shall be made in the Cargo Record Book;

(d) if the tank is to be further cleaned and emptied at sea, the Master shall:

(i) ensure that the approved procedures referred to in Sub-Paragraph 2 c) of Paragraph D of this

Regulation are complied with and that the appropriate entries are made in the Cargo Record Book; and

(ii) ensure that any discharge into the sea is made in accordance with the requirements of Sub-Paragraph 2 of Paragraph D of this Regulation and an appropriate entry is made in the Cargo Record Book;

(e) if after unloading a Category B substance, any residues of tank washings are to be retained on board until the ship is outside the Baltic Sea Area, the Master shall so indicate by an appropriate entry in the Cargo Record Book.

Category C Substances

Subject to such surveillance and approval by the authorized or appointed surveyor as may be deemed necessary by the Contracting Party, the Master of ship shall, with respect to a 'Category C Substance, ensure compliance with the following:

(a) if a tank is partially unloaded or unloaded but not cleaned, an appropriate entry shall be made in the Cargo Record Book;

(b) if the tank is to be cleaned at ses;

  (i) the cargo piping system serving that tank shall be drained and an appropriate entry made in the Cargo Record Book;

  (ii) the quantity of substances remaining in the tank shall not exceed the maximum quantity which may be discharged into the sea for that substance under Sub-Paragraph 3 c) of Paragraph D of this Regulation. An appropriate entry shall be made in the Cargo Record Book

  (iii) where it is intended to discharge the quantity of substance remaining into the sea the approved procedures shall be complied with and the necessary dilution of the substance satisfactory for such discharge shall be achieved An appropriate entry shall be made in the Cargo Record Book; or

(iv) where the tank washings are not discharged into the sea, if any internal transfer of tank washings takes place from that tank an appropriate entry shall be made in the Cargo Record Book; and

(v) any subsequent discharge into the sea of such tank washings shall be made in accordance with the requirements of Sub-Paragraph 3 of Paragraph D of this Regulation;

(c) if the tank is to be cleaned in port:

(i) the tank washings shall be discharged to reception facility and an appropriate entry shall be made in the Cargo Record Book; or

(ii) the tank washings shall be retained on board the ship and an appropriate entry shall be made in the Cargo Record Book indicating the location and disposition of the tank washings;

(d) if after unloading a Category C substance within the Baltic Sea Area~ any residues or tank washings are to be retained on board until the ship is outside the area, the Master shall so indicate by an appropriate entry in the, Cargo Record Book.

Category D Substances

7. The master of a ship shall, with respect to a Category D Substance, ensure compliance with the following:

(a) if a tank is paritally unloaded or unloaded but not cleaned, and appropriate entry shall be made in the Cargo Record Book;

(b) if the tank is to be cleaned at sea:

(i) the cargo piping system serving that tank shall be drained and an appropriate entry made in the Cargo Record Book;

(ii) where it is intended to discharge the quantity of substance remaining into the sea necessary dilution of the substance satisfactory for such a discharge shall be achieved. An appropriate entry shall be made in the Cargo Record Book;

(iii) where the tank washings are not discharged into the sea, if any internal transfer of tank washings takes place from that tank an appropriate entry shall be made in the Cargo Record Book; and

(iv) any subsequent discharge into the sea of such tank washings shall be made in accordance with the requirements of Sub-Paragraph 4 of Paragraph D of this Regulation;

(c) if the tank is to be cleaned in port:

(i) the tank washings shall be discharged to a reception facility and an appropriate entry shall be made in the Cargo Record Book; or

(ii) the tank washings shall be retained on board the ship and an appropriate entry shall be made in the Cargo Record Book indicating the location and disposition of the tank washings.

Discharge from a Slop Tank

8. Any residues retained on board in a slop tank, including those from pump room bilges, which contain a Category A or a Category 13 substance, shall be discharged to reception facility in accordance with the provisions of Sub-Paragraph 1 or 2 of Paragraph D of this Regulation, whichever is applicable. An appropriate entry shall be made in the Cargo Record Book.

9. Any residues retained on board in a slop tank, including those from pump room bliges, which contain a Category C substance in excess of the aggregate of the maximum quantities specified in Sub-Paragraph 3 c) of Paragraph D of this Regulation shall be discharged to reception facility. An appropriate entry shall be made in the Cargo Record book.

G Cargo Record Book

1. Every ship to which this Regulation applies shall be provided with a Cargo Record, whether as part of the ship's official log book or otherwise, in the form recommended by the Commission.
2. The Cargo Record Book shall be completed, on a tank-to-tank basis, whenever any of the following operations with respect to a noxious liquid substance takes place in the ship:
   (i) Loading of cargo (substances of Category A,B,C and D);
   (ii) international transfer of cargo;, (iii) unloading of cargo;
   (iv) cleaning of cargo tanks;
   (v) ballasting of cargo tanks;
   (vi) discharge of ballast from cargo tanks; (vii) disposal of residues;
   (viii) discharge onto the sea in accordance with

Paragraph D of this Regulation

3. In the event of any discharge of the kind referred to in Annex VI of the present convention and Paragraph E of this Regulation of any noxious liquid substance of mixture containing such substance, whether intentional or accidental, an entry shall be made in the Cargo Record Book stating the circumstances of, and the reasons for, the discharge.

4. When a surveyor appointed or authorized by a Contracting Party to supervise any operations under this Regulation- has inspected a ship, then that surveyor shall make an appropriate entry in the Cargo Record Book.

5. Each operation referred to in Sub-Paragraphs 2 and 3 of this Paragraph shall be fully recorded without delay in the Cargo Record Book so that all the entries in the Book appropriate to that operation are completed. Each entry shall be signed by the officer or officers in charge of the operation concerned and, when the ship is manned, each page shall be signed by the Master of the ship. The entries in the Cargo Record Book shall be in an official language of the State whose flag the ship is entitled to fly, and, except when the ship is engaged in domestic voyages, in English or French. The entries in an official national language of the State whose flag the ship is entitled to fly shall prevail in case of dispute or discrepancy.

6. The Cargo Record Book shall be kept in such a place as to be readily available for inspection and, except in the case of unmanned ships under tow, shall be kept on board the ship. It shall be retained for period of three years after the last entry has been made.

7. The competent authority of a Contracting Party may inspect the Cargo Record Book on board any ship to which this Regulation applies while the ship is in its port, and may make a copy of any entry in that Book and may require the Master of the ship to certify that the copy is a true copy of such entry. Any copy so made which has been certified by the Master of the ship as a true copy of an entry in the ship's Cargo Record Book shall be made admissible in any judicial proceeding as evidence of the facts

stated in the entry. The inspection of a Cargo Record Book and the taking of Certified copy by the competent authority under this Paragraph shall be performed as expeditiously as possible without causing the ship to be unduly delayed.

Reception Facilities

1. The Contracting Parties undertake to ensure the provision of reception facilities according to the needs of ships using their ports, terminals or repair ports of the Baltic Sea Area as follows:
   (a) cargo loading and unloading ports and terminals shall have facilities adequate for reception without undue delay to ships of such residues and mixtures containing noxious liquid substances as would remain for disposal from ships carrying them as a consequence of the application of this Regulation; and
   (b) ship repair ports undertaking repairs to chemical tankers shall have facilities adequate for the reception of residues and mixtures containing noxious liquid substances.
2. Each Contracting Party shall determine the types of facilities provided for the purpose of Sub-Paragraph 1 of this Paragraph at its cargo loading and unloading ports, terminals and ship repair ports of the Baltic Sea Area.

Regulation 6

Harmful substances in packaged forms

A. The Contracting Parties shall as soon as possible apply suitable uniform rules for the carriage of harmful substances in packaged forms or in freight containers, portable tanks or road and rail tank wagons.

B. With respect to certain harmful substances, as may

be designated by the Commission, the Master or owner of the ship or his representative shall notify the appropriate port authority of the intent to load or unload such substances at least 24 hours prior to such action.

C. A report of an incident involving harmful substances shad be made in accordance with the provisions of Annex VI of the present Convention.

Regulation 7

Sewage

The Contracting Parties shall apply the provisions of Paragraphs A to D of this Regulation on discharge of sewage ships while operating in the Baltic Sea Area.

Definitions

For the purposes of this Regulation:

1. "New ship" means a ship:
   (a) for which the building contract is placed, or in the absence of a building contract, the keel of which is laid or which is at a similar stage of construction, on or after the date of entry into force of the present Convention; or
   (b) the delivery of which is three years or more after the date of entry into force of the present Convention;
2. "Existing ship" means a ship which is not a new ship.
3. "Sewage" means:
   (a) damage and other wastes from any of toilets urinals, and WC scuppers;
   (b) drainage from medical premises (dispensary, sick bay, etc.) via wash basins, wash tubs and scuppers located in such premises;

(c) drainage from spaces containing living animals; or

(d) other waste waters when mixed with the damages defined above.

4. "Holding tank" means a tank used for the collection and storage of sewage.

Application

1. The provisions of this Regulation shall apply to:

   (a) new ships certified to carry more than 100 persons from a date not later than 1 January 1977;

   (b) existing ships certified to carry more than 400 persons form a date not later than 1 January 1978; and

   (c) other ships, as specified in Sub-Paragraph (i), (ii), and (iii), from dates decided by the Contracting Parties on recommendation by the Commission:

      (i) ships of 200 tons gross tonnage and above;

      (ii) ships of less than 200 tons gross tonnage which are certified to carry more than 10 persons;

      (iii) ships which do not have measured gross tonnage and are certified to carry more than 40 persons.

In the case of new such ships the date shall be not later than 1 January 1979. In the case of existing such ships the date shall be not later than ten years after the date decided for new ships.

2. A Contracting Party may, if it is satisfied that the application of the provisions of Sub-Paragraph 1 b) of this Paragraph with respect to certain ship would necessitate constructional alterations which would be unreasonable exempt the ship from the

application until a date not later than ten years after the date of entry into force of the present Convention.

Discharge of Sewage

1. Subject to the provisions of Paragraph D of this Regulation, the discharge of sewage into the sea is prohibited, except when:

   (a) the ship is discharging comminuted `and disinfected sewage using a system approved by the Administration at a distance of more than 4 nautical miles form the nearest land, or sewage which is not comminuted or disinfected at a distance of more than 12 nautical miles form the nearest land, provided that any case the sewage that has been stored in holding tanks shall not be discharged instantaneously but, at a moderate rate when the ship is en route and proceeding at not less than 4 knots; or

   (b) the ship has in operation a sewage treatment plant which has been approved by the Administration, and

      (i) the test results of the plant are laid down in a document carried by the ship;

      (ii) additionally, the effluent shall not produce visible floating solids, nor cause discolouration of the surrounding water; or

   (c) the ship is situated in the waters under the jurisdiction of a State and is discharging sewage in accordance with such less stringent requirements as may be imposed by such State.

2. When the sewage is mixed with wastes or waste water having different discharge requirements, more stringent requirements shall apply.

Exceptions

Paragraph C of this Regulation shall not apply to:

(a) the discharge of sewage from a ship necessary for the purpose of securing the safety of ship and those on board or saving life at sea; or

(b) the discharge of sewage resulting from damage to a ship or its equipment if all reasonable precautions have been taken before and after the occurrence of the damage for the purpose or preventing or minimizing the discharge.

Reception Facilities

1. Each Contracting Party undertakes to ensure the provision of facilities at its ports and terminals of the Baltic Sea Area for the reception of sewage.

   Without causing undue delay to ships, adequate to meet the needs of the ships using them.

2. To enable pipes of reception facilities to be connected with the ship's discharge pipeline, both lines shall be fitted with a standard discharge connection in accordance .With the following table:

Standard Dimensions of Flanges for Discharge Connections

Description	Dimension
Out side diameter	210 mm
Inner diameter	According to pipe outside diameter
Bolt circle diameter	170 mm
Slots in flange diameter equidistantly	4 holes 18 mm in placed on a bolt circle of the above diameter, slotted to the flange periphery. The slot width to be 18 mm
Flange thickness	16 mm
Bolts and nuts; diameter and of	4 each of 16 mm in
quantity and diameter	suitable length

The flange is designed to accept pipes up to a maximum internal diameter of 100 mm and shall be of steel or other equivalent material having a flat face. This flange, together with a suitable gasket, shall be suitable for service pressure of 6 kg/cm^2.

For ships having a moulded depth or 5 meters and less, the inner diameter of the discharge connection may be 38 millimeters.

Regulation 8

Garbage

The Contracting Parties shall as soon as possible but not later than 1 January 1976 or on the date of entry into force of the present Convention, whichever occurs later, apply the provisions of Paragraphs A to D of this Regulation on the disposal of garbage from ships while operating in the Baltic Sea Area.

Definition

For the purposes of this Regulation:

"Garbage" means all kinds of victual, domestic and operational waste excluding fresh fish and parts thereof, generated during the normal operation of the ship and liable to be disposed of continuously or periodically except those substances which are defined or listed in other Regulations of this Annex.

Disposal of Garbage

1. Subject to the provisions of Paragraphs C and D of this Regulation:
   (a) disposal into the sea of the following is prohibited:
      (i) all plastics, including but not limited to synthetic ropes, synthetic fishing nets and plastic garbage bags; and

(ii) all other garbage, including paper products, rags, glass, metal, bottles, crockery, lining and packing materials;

(b) disposal into the sea of food wastes shall be made as far as practicable from land , but in-any case not less than 12 nautical miles from the nearest land,

2. When the garbage is mixed with other discharges having different disposal or discharge requirements the more stringent requirements shall apply.

Special Requirements for Fixed and Floating Platforms

1. Subject to the provision of Sub-paragraph 2 of this paragraph, the disposal of any materials regulated by this Regulation is prohibited from fixed or floating platforms engaged in the exploration, exploitation and associated off-shore processing of sea-bed mineral resources, and all other ships when alongside or within 500 meteres of such platforms.

2. The disposal into the sea of food wastes may be permitted when they have passed through a comminuter or grinder from such fixed or floating platforms located more than 12 nautical miles from land and all other ships when alongside or within 500 meters of such platforms. Such comminuted or ground food wastes shall be capable of passing through a screen with openings no greater than 25 millimeters.

Exceptions

Paragraphs B and C of this Regulation shall not apply to:

(a) the disposal of garbage from a ship necessary for the purpose of securing the safety of ship and those on board or saving life at sea; or

(b) the escape of garbage resulting from damage to a ship or its equipment provided all reasonable precautions have been taken before and after the occurrence of the damage, for the purpose of preventing or minimizing the escape; or

(c) the accidental loss of synthetic fishing nets or synthetic material incidental to the repair of such nets, provided that all reasonable precautions have been taken to prevent such loss.

Reception Facilities

Each Contracting Party undertakes to ensure the provision of facilities at its ports and terminals of the Baltic Sea Area for the reception of garbage, without causing undue delay to ships, and according to the needs of the ships using them.

Annex V

Exceptions from the General Prohibition of Dumping of Waste and other Matter in the Baltic Sea Area

Regulation I

In accordance with Paragraph 2 of Article 9 of the present Convention the prohibition of dumping shall not apply to the disposal at sea of dredged spoils provided that:

1. They do not contain significant quantities and concentrations of substances to be defined by the Commission and listed in Annexes I and II of the present Convention; and
2. The dumping is carried out under a prior special permit given by the appropriate national authority, either
   (a) within the area of the territorial sea of the Contracting Party; or
   (b) outside the area of the territorial sea, whenever necessary, after prior consultations in the Commission.

When issuing such permits the Contracting Party shall comply with the provisions in Regulation 3 of this Annex.

Regulation 2

1. The appropriate national authority referred to in Paragraph 2 of Article 9 of the present Convention shall:
   (a) issue special permits provided for the Regulation 1 of this Annex;
   (b) keep records of the nature and quantities of matter permitted to be dumped and the location, time and method of dumping;
   (c) collect available information concerning the nature and quantities of matter that has been dumped in the Baltic Sea Area recently and up to the coming in to force of the present Convention, provided that the dumped matter in question could be liable to contaminate water or organisms in the Baltic Sea Area, to be caught by fishing equipment, or otherwise to give rise to harm, and the location, time and method of such dumping.
2. The appropriate national authority shall issue special permits in accordance with Regulation 1 of this Annexd inrespect of matter intended for dumping in the Baltic Sea Area:
   (a) loaded in its territory;
   (b) loaded by a vessel or aircraft registered in its territory or flying its flag, when the loading occurs in the territory of a State not Party to the present Convention,.
3. When issuing permits under Sub-Paragraph 1 a) and above, the appropriate national authority shall comply with Regulation 3 of this Annex, together

with such additional criteria, measures and requirements as they may consider relevant.

4. Each Contracting Party shall report to the Commission, and where appropriate to other. Contracting parties, the information specified in Sub-Paragraph 1 c) of Regulation 2 of this Annex. The procedure to be followed and the nature of such reports shall be determined by the Commission.

***Regulation* 3**

When issuing special permits according to Regulation 1 of this Annex the appropriate national authority shall take into account:

1. Quantity of dredged spoils to be dumped.
2. The content of the matter referred to in Annex I and II of the present Convention.
3. Location (e.g. co-ordinates of the dumping area, depth and distance from coast) and its relation to areas of special interest (e.g. amenity areas, spawrning, nursery and fishing areas, etc.)
4. Water characteristics, if dumping is carried outside the territorial sea, consisting of:
   (a) hydrographic properties (e.g. temperature, salinity density, profile);
   (b) chemical properties (e.g. pH dissolved Oxygen nutrients);
   (c) biological properties (e.g. primary production and benthic animals).

The data should include sufficient information oil the annual mean levels and the seasonal variation *Of* the properties mentioned in this paragraph.

5. The existence and effects of other dumping which may have been carried out in the dumping area.

Regulation 4

Reports made in accordance with Paragraph 5 of Article 9 of the present Convention shall include the following information:

1. Location of dumping, characteristics of dumped material, and counter measures taken:
   (a) location (e.g. coordinates if the accidental dumping site, depth and distance from the coast);
   (b) method of deposit; .
   (c) quantity and composition of dumped matter as well as its physical (e.g. solubility and density), chemical and biochemical (e.g. oxygen demand, nutrients) and biological properties (e.g. presence of viruses, bacteria yeasts parasites);
   (d) toxicity;
   (e) content of the substances referred to in Annexes I and II of the present Convention;
   (t) dispersal characteristics (e.g. effects of currents and wind, and horizontal transport and vertical mixing);
   (g) water characteristics (e.g. temperature, pH, redox conditions, salinity and stratification);
   (h) bottom characteristics (e.g. topography, geological characteristics and redox conditions);
   (i) counter measures taken and follow-up operations carried out or planned.
2. General consideration and conditions:
   (a) possible effect on amenities (e.g. floating or stranded material, turbidity, objectionable odour, discolouration and foaming);

(b) possible effect on marine life, fish and shellfish culture, fish stocks and fisheries, seaweed harvesting and cultures; and

(c) possible effects on other uses of the sea (e.g. impairment of water quality for industrial use, underwater corrosion of structures, interference with ship operations from floating materials, interference with fishing or navigation and protection of areas of special importance for scientific or conservation purposes).

Annex VI

Co-operation in combatting Marine Pollution

Regulation 1

For the purposes of this Annex:

1. "Ship" means a vessel of any type whatever operating in the marine environment and includes hydofoil boats, air-cushion vehicles, submersibles, floating craft and fixed or floating platforms.

2. "Administration" means the Government of the State under whose authority the ship is operating. With respect to a ship entitled to fly a flag of any State, the Administration is the Government of that State. With respect to fixed or floating platforms engaged in exploration and exploitation of the seabed and subsoil thereof adjacent to the coast over which the coastal State exercises sovereign rights for the purposes of exploration and exploitation of their natural resources, the Administration is the Government of the coastal State concerned

3. (a) "Discharge", in relation to harmful substances or effluents containing such substances, means any release howsoever caused from a ship and includes any escape, disposal, spilling, leaking, pumping, emitting or emptying.

(b) "Discharge" does not include:

(i) dumping within the meaning of the Convention on the Prevention of Marine Pollution by Dumping of Wastes and Other Matter done at London on 29 December 1972; or

(ii) release of harmful substances directly arising from the exploration and associated offshore processing of seabed mineral resources; or

(iii) release of harmful substances for purposes of legitimate scientific research into pollution abatement or control.

Regulation 2

The Contracting Parties undertake to maintain ability to combat spillages of oil and other harmful substances on the sea. This ability shall include adequate equipment, ships and manpower prepared for operations in coastal waters as well as on the high sea.

Regulation 3

The Contracting Parties shall, without prejudice to Paragraph 4 of Article 4 of the present Convention, develop and apply, individually or in cooperation, surveillance activities covering the Baltic Sea Area, inorder to spot and monitor oil and other harmful substances released into the sea.

Regulation 4

In the case of loss overboard of harmful substances in packages, freight containers, portable tanks, or road and rail tank wagons, the Contracting Parties shall cooperate in the salvage and recovery of such packages, containers or tanks so as to minimize the danger to the environment.

Regulation 5

1. The Contracting Parties, also being parties to the International Convention for the Prevention of Pollution from ships, 1973, as modified by the Protocol of 1978 relating thereto (MARPOL73/78) apply in conformity with that protocol I to MARPOL 73/78 on reports on incidents involving harmful substances. These provisions shall also be applied with regard to significant spillagers, of oil or other harmful substances in cases not covered by Article 8 of MARPOL 73/78.
2. The Contracting Parties shall request masters of ships and pilots of aircraft to report without delay in accordance with this system on significant spillages of oil or other harmful substances observed at sea. Such reports should as far as possible contain the following data: time, position, wind and sea conditions, and kind, extent and porbable source of the spill observed.
3. The master of ship involved in an incident referred to in Paragraph 1 of this regulation, or other person having charge of the ship shall without delay and t~ the fullest extent possible report in accordance with this system and with the provisions of the Appendix to the present Annex.
4. Each Contracting Party undertakes to issue instructions to its maritime inspection vessels and aircraft and to other appropriate services to report to its authorities any observation or incident referred to in Paragraph 1 of this Regulation. Such reports shall as far as possible contain the data referred to in paragraphs 2 or 3 of this Regulation respectively, as well as possible indications on the spreading or drifting tendencies of the spill question.
5. Whenever a Contracting Party is aware of a casuality or the presence of spillages of oil or other

harmful substances in the baltic Sea Area likely to constitute a serious threat to the marine environment of the Baltic Sea Area or the coast or related interests of any other Contracting Party, it shall without delay transmit all revelant information thereon to the Contracting Party which may be affected by the pollutant and, as regards ship casualty incidents, to the Administration of the ship involved.

Regulation 6

Each Contracting Party shall request masters of ships flying its flag to provide, in case of an incident, on request by the proper authorities, such detailed information about the ship and its cargo which is relevant to actions for preventing or combatting pollution of the sea, and to cooperate with these authorities.

Regulation 7

1. (a) The contracting Parties shall as soon as possible agree bilaterally or multilaterally on those regions of the Baltic Sea Area in which they will take action for combatting or salvage activities whenever a significant spillage of oil or other harmful substances or any incidents causing or likely to cause pollution within the Baltic Sea Area have occurred or are likely to occur. Such agreements shall not prejudice any other agreements 'concluded between Contracting Parties concerning the same subject. The neighbouring States shall ensure the harmonization of the different agreements. The Contracting Parties shall inform each other about such agreements.

   The contracting Parties may ask the Commission for assistance to reach agreement, if needed.

   (b) The Contracting Party within whose region a situation as described in Regulation 1 of this Annex occurs shall make the necessary assessments of the situation and take adequate action in order to avoid or minimize subsequent pollution effects and shall keep drifting parts of the spillage under observation until no further action is called for,

2. In the case that such a spillage is drifting or is likely to drift into a region, where another Contracting Party should take action for purposes as definded in Subparagraph 1 a) of this regulation, that Party shall without delay be informed of the situation and the actions that have been taken.

Regulation 8

A Contracting Party requiring assistance for combatting spillages of oil or other harmful substances .at sea is entitled to call for assistance by other. Contracting Parties, starting with those who seem likely also to be affected by the spillage. Contracting Parties called upon for assistance in accordance with this Regulation shall use their best endeavours to bring such assistance.

Regulation 9

1. The Contracting Parties shall provide information to the other Contracting Parties and the Commission about
   (a) their national organization for dealing with spillages at sea of oil and other harmful substances;
   (b) national regulations and other matters which have a direct bearing on combatting pollution at sea by oil and other harmful substances; .
   (c) the competent authority responsible for receiving and dispatching reports of pollution at sea by oil and other harmful substances;
   (d) the competent authorities for dealing with questions' concerning measures of mutual assistance, information and cooperation

between the Contracting Parties according to this Annex;

(e) actions taken in accordance with Regulation 8 of this Annex.

2. The Contracting Parties shall exchange information of research and development programs and results concerning ways in which pollution by oil and other harmful substances at sea may be dealt with and experiences in combatting such pollution.

Regulation 10

The authorities referred to in Sub-Paragraph 1 d) of Regulation 9 of this Annex shall establish direct contact and cooperate in operational matters.

5 Paris Convention for the Prevention of Marine Pollution (1974)

The contracting parties

Recognizing that the marine environment and the fauna and flora which it supports are of vital importance to all nations;

Mindful that the ecological equilibrium and the legitimate uses of the sea are increasingly threatened by pollution;

Considering the recommendations of the United Nations conference on the human environment, held in Stockholm in June 1972;

Recognizing that concerted action at national, regional and global levels is essential to prevent and combat marine pollution;

Convinced that international action to control the pollution of the sea from land-based sources can and should be taken without delay, as part of progressive and coherent measures to protect the marine environment from pollution, whatever its origin, including current efforts to combat the pollution of international waterways;

Considering that the common interests of States concerned with the same marine area should induce them to cooperate at regional or subregional levels;

Recalling the convention for the prevention of marine pollution by dumping from ships and aircraft concluded in Oslo on February 15, 1972,

Have Agreed as follows:

Article 1

1. The Contracting Parties pledge themselves to take all possible steps to prevent pollution of the sea, by which is meant the introduction by man, directly or indirectly, of substances or energy into the marine environment (including estuaries) resulting in such deleterious effects as hazards to human health, harm to living resources and to marine ecosystems, damage to amenities or interference with other legitimate uses of the sea.
2. The Contracting Parties shall adopt individually and jointly measures to combat marine pollution from land-based sources in accordance with the provisions of the present convention and shall harmonize their policies in this regard.

Article 2

The present convention shall apply to the maritime area within the following limits:

(a) those parts of the Atlantic and Arctic Oceans and the dependent seas which lie north of 36° north latitude and between 42° west longitude and 51° east longitude, but excluding:

(i) the Baltic Sea and Belts lying to the -south and east of lines drawn from Hasenore Head to Gniben Point, from Korshage to Spodsbjerg and from Gilbjerg Head to Kullen and

(ii) the Mediterranean Sea and its dependent seas as far as the point of intersection of the parallel of 36° north latitude and the meridian of 5° 36'west longitude;

(b) that part of the Atlantic Ocean north of 59° north latitude and between 44° west longitude and 42° west longitude.

Article 3

For the purpose of the present convention:

(a) maritime area means: the high seas, the territorial seas of Contracting Parties and waters on the landward side of the base lines from which the breadth of the territorial sea is measured and extending in the case of watercourses, unless otherwise decided under Article 16(c) of the present convention, up to the freshwater limit;

(b) 'Freshwater limit' means: the place in the watercourse where, at low tide and in a period of low freshwater flow, there is an appreciable increase in salinity due to the presence of seawater;

(c) 'pollution from land-based sources' means: the pollution of the maritime area

(i) through watercourses,

(ii) from the coast, including introduction through underwater or other pipelines,

(iii) from man-made structures placed under the jurisdiction of a Contracting Party within the limits of the area to which the present convention applies.

(iv) by emissions into the atmosphere from land or from man-made structures as defined in subparagraph (iii) above.

Article 4

1. The Contracting Parties undertake:
   (a) to eliminate, if necessary by stages, pollution of the maritime area from land-based sources by substances listed in Part I of Annex A to the present convention;
   (b) to limit strictly pollution of the maritime area from land-based sources by substances listed in Part II of Annex A to the present convention.
2. In order to carry out the undertakings in paragraph 1 of this Article, the Contracting Parties, jointly or individually as appropriate, shall implement programmes and measures:
   (a) for the elimination, as a matter of urgency, of pollution of the maritime area from land-based sources by substances listed in Part I of Annex A to the present convention;
   (b) for the reduction or, as appropriate, elimination of pollution of the maritime area from land-based sources by substances listed in Part II of Annex A to the present convention. These substances shall be discharged only after approval has been granted by the appropriate authorities within each Contracting State. Such approval shall be periodically reviewed.
3. The programmes and measures adopted under paragraph 2 of this Article shall include, as appropriate, specific regulations or standards governing the quality of the environment, discharges into the maritime area, such discharges into watercourses and emissions into the atmosphere as affect the maritime area, and the composition and use of substances and products. These programmes and measures shall take into account the latest technical developments.

The programmes shall contain time limits for their completion.

4. The Contracting Parties may, furthermore, jointly or individually as appropriate, implement programmes or measures to forestall, reduce or eliminate pollution of the maritime area from land-based sources by a substance riot then listed in Annex A to the present convention, if scientific evidence has established that a serious hazard may be created in the maritime area by that substance and if urgent action is necessary.

Article 5

1. The Contracting Parties undertake to adopt measures to forestall and, as appropriate, eliminate pollution of the maritime area from land-based sources by radioactive substances referred to in Part III of Annex A of the present convention.

2. Without prejudice to their obligations under other treaties and conventions, in implementing this undertaking the Contracting Parties shall:

   (a) take full account of the recommendations of the appropriate international organizations and agencies;

   (b) take account of the monitoring procedures recommended by these international organizations and agencies;

   (c) coordinate their monitoring and study of radioactive substances in accordance with Articles 10 and 11 of the present convention.

Article 6

1. With a view to preserving and enhancing the quality of the marine environment, the Contracting Parties, without prejudice to the provisions of Article 4, shall endeavour:

(a) to reduce existing pollution from land-based sources;

(b) to forestall any new pollution from land-based sources, including that which derives from new substances.

2. In implementing this undertaking, the Contracting Parties shall take account of.

(a) the nature and quantities of the pollutants under consideration;

(b) the level of existing pollution;

(c) the quality and obsorptive capacity of the receiving waters of the maritime area;

(d) the need for an integrated planning policy consistent with the requirement of environmental protection.

Article 7

The Contracting Parties agree to apply the measures

— they adopt in such a way as to avoid increasing pollution:

— in the seas outside the area to which the present convention applies;

— in the maritime area covered by the present convention, originating otherwise than from land-based sources.

Article 8

No provision of the present convention shall be interpreted as preventing the Contracting Parties from taking more stringent measures to combat marine pollution from land-based sources.

Article 9

1. When pollution from land-based sources originating from the territory of a Contracting Party by substances not listed in Part I of Annex A to the

present convention is likely to prejudice the interests of one or more of the other parties to the present convention, the Contracting Parties concerned undertake to enter into consultation, at the request of any one of them, with a view to negotiating a cooperation agreement.

At the request of any Contracting Party concerned, the Commission referred to in Article 15 of the present convention shall consider the question and may make recommendations with a view to reaching a satisfactory solution.

The special agreements specified in paragraph 1 of this Article may, among other things, define the areas to which they shall apply, the quality objectives to be achieved, and the methods for achieving these objectives including methods for the application of appropriate standards and the scientific and technical information to be collected.

The Contracting Parties signatory to these special agreements shall, through the medium of the Commission, inform the other Contracting Parties of their purport and of the progress made in putting them into effect.

Article 10

The Contracting Parties agree to establish complementary or joint programmes of scientific and technical research, including research into the best methods of eliminating or replacing noxious substances so as to reduce marine pollution from land-based sources, and to transmit to each other the information so obtained. In doing so they shall have regard to the work carried out, in these fields, by the appropriate international organizations and agencies.

Article 11

The Contracting Parties agree to set up progressively and

to operate within the area covered by the present convention a permanent monitoring system allowing:

- the earliest possible assessment of the existing level of marine pollution;
- the assessment of the effectiveness of measures for the reduction of marine pollution from land-based sources taken under the terms of the present conventions.

For this purpose the Contracting Parties shall lay down the ways and means of pursuing individually or jointly systematic and ad hoe monitoring programmes. These programmes shall take into account the deployment of research vessels and other facilities in the monitoring area.

The programmes shall take into account similar programmes pursued in accordance with conventions already in force and by the appropriate international organizations and agencies.

Article 12

1. Each Contracting Party undertakes to ensure compliance with the provisions of this convention and to take in its territory appropriate measures to prevent and punish conduct in contravention of the provisions of the present convention.
2. The Contracting Parties shall inform the Commission of the legislative·and administrative measures they have taken to implement the provisions of the preceding paragraph.

Article 13

The Contracting Parties undertake to assist one another as appropriate to prevent incidents which may result in pollution from land-based sources, to minimize and eliminate the consequences of such incidents, and to exchange information to that end.

Article 14

1. The provisions of the present convention may not be invoked against a Contracting Party to the extent that the latter is prevented, as a result of pollution having its origin in the territory of a non-contracting State, from ensuring their full application.
2. However, the said Contracting Party shall endeavour to cooperate with the non-contracting States so as make possible the full application of the present convention.

Article 15

A Commission composed of representatives of each of the Contracting Parties is hereby established. The Commission shall meet at regular intervals and at any time when due to special circumstances it is so decided in accordance with its rules of procedure.

Article 16

It shall be the duty of the Commission:

(a) to exercise overall supervision over the implementation of the present convention;

(b) to review generally the condition of the seas within the area to which the present convention applies, the effectiveness of the control measures being adopted and the need for any additional or different measures;

(c) to fix, if necessary, on the proposal of the Contracting Party or Parties bordering on the same watercourse and following a standard procedure, the limit to which the maritime area shall extend in that watercourse;

(d) to examine the feasibility of and, as appropriate, to draw up, in accordance with Article 4 of the present convention, programmes and measures for the elimination or reduction of pollution from land-based sources;

(e) to make recommendations in accordance with the provisions of Article 9;

(f) to receive and review information and distribute it to the Contracting Parties in accordance with the provisions of Articles 11, 12 and 17 of the present convention;

(g) to make, in accordance with Article 18, recommendations regarding any amendment to the lists of substances included in Annex A to the present convention;

(h) to discharge such other functions, as may be appropriate, under the terms of the present convention.

Article 17

The Contracting Parties, in accordance with a standard procedure, shall transmit to the Commission:

(a) the results of monitoring pursuant to Article 11;

(b) the most detailed information available on the substances listed in the Annexes to the present convention and liable to find their way into the maritime area.

The Contracting Parties shall endeavour to improve progressively techniques for gathering such information which can contribute to the revision of the pollution reduction programmes drawn up in accordance with Article 4 of the present convention.

Article 18

1. The Commission shall draw up its own Rules, of Procedure which shall be adopted by unanimous vote.
2. The Commission shall draw up its own Financial Regulations which shall be adopted by unanimous vote.

3. The Commission shall adopt, by unanimous vote, programmes and measures for the reduction or elimination of pollution from land-based sources as provided for in Article 4, programmes for scientific research and monitoring as provided for in Articles 10 and 11, and decisions under Article 16(c).

The programmes and measures shall commence for and be-applied by all Contracting Parties 200 days after their adoption, unless the Commission specifies another date.

Should unanimity not be attainable, the Commission may nonetheless adopt a programme or measures by a three-quarters majority vote of its members. The programmes or measures shall commence for those Contracting Parties which voted for them 200 days after their adoption, unless the Commission specifies another date, and for any other Contracting Party after it has explicitly accepted the programme or measures, which it may do at any time.

4. The Commission may adopt recommendations for amendments to Annex A to the present convention by a three-quarters majority vote of its members and shall submit them for the approval of the Governments of the contracting Parties. Any Government of a Contracting Party that is unable to approve an amendment shall notify the depositary Government in writing within a period of 200 days after the adoption of the recommendation of amendment in the Commission. Should no such notification be received, the amendment shall enter into force for all Contracting Parties 230 days after the vote in the Commission. The depositary Government shall notify the Contracting Parties as soon as possible of the receipt of any notification.

Article 19

Within the area of its competence, the European

Economic Community is entitled to a number of votes equal to the number of its Member States which are Contracting Parties to the present convention.

The European Economic Community shall not exercise its right to vote in cases where its Member States exercise theirs and conversely.

Article 20

The depositary Government shall convene the first meeting of the Commission as soon as possible after the coming into force for the present convention.

Article 21

Any dispute between Contracting Parties relating to the interpretation or application of the present convention, which cannot be settled otherwise by the parties concerned, for instance by means of inquiry or conciliation within the Commission, shall, at the request of any of those parties, be submitted to arbitration under the conditions laid down in Annex B to the present convention.

Article 22

The present convention shall be open for signature at Paris from June 4, 1974 to June 30, 1975, by the States invited to the diplomatic conference on the convention for the prevention of marine pollution from land-based sources, held at Paris, and by the European Economic Community.

Article 23

The present convention shall be subject to ratification, Acceptance or approval. The instruments of ratification, acceptance or approval shall be deposited with the Government of the French Republic.

Article 24

1. After June 30, 1975, the present convention shall b e open for accession by States referred to in Article 22 and by European Economic Community.

2. The present convention shall also be open for accession from the same date by any other Contracting Party to the convention for the prevention of marine pollution by dumping from ships and aircraft, opened for signature at Oslo on 15th February 1972.
3. From the date of its entry into force, the present convention shall be open for accession by any States not referred to in Article 22, located upstream on watercourses crossing the territory of one or more Contracting Parties to the present convention and reaching the maritime area defined in Article 2.
4. The Contracting Parties may unanimously invite other States to accede to the present convention. In that case the maritime area in Article 2 may, if necessary, be amended in accordance with Article 27 of the present convention.
5. The instruments of accession shall be deposited with the Government of the French Republic.

Article 25

1. The present convention shall come into force on the thirtieth day following the date of deposit of the seventh instrument of ratification, acceptance, approval or accession.
2. For each Party ratifying, accepting or approving the present convention or acceding to it after the deposit of the seventh instrument of ratification, acceptance, approval or accession, the present convention shall enter into force on the thirtieth day after the date of deposit by the party of its instrument of ratification, acceptance, approval or accession.

Article 26

At any time after the expiry of two years from the date of coming into force of the present convention in relation to any Contracting Party such party may withdraw from the convention by notice in writing to the depositary Government. Such notice shall take effect one year after the date on which it is received.

Article 27

1. The depositary Government shall, at the request of the Commission on a decision taken by a two-thirds majority of its members, call a conference for the purpose of revising or amending the present convention.
2. Upon accession by a State as provided for in paragraphs 2, 3 and 4 of Article 24, the maritime area in Article 2 may be amended upon a proposal by the Commission adopted by a unanimous vote. These amendments shall enter into force after unanimous approval by the Contracting Parties.

Article 28

The depositary Government shall inform the Contracting Parties's and those referred to in Article 22:

(a) of signatures to the present convention of the deposits of instruments of ratification, acceptance, approval or accession, and of notices of withdrawal in accordance with Articles 22, 23, 24 and 26;,

(b) of the date of which the present convention comes into force in accordance with Article 25;

(c) of the receipt of notifications of approval or objection, and of the entry into force of amendments to the present convention and its Annexes, in accordance with Article 18 and 27.

Article 29

The original of the present convention of which the French and English texts shall be equally authentic, shall be deposited with the government of the French Republic which shall send certified copies thereof to the Contracting Parties and the States referred to in Article 22 and shall deposit a certified copy with the Secretary General of the United Nations for registration and publication in accordance with Article 102 of the United Nations Charter.

In Witness Whereof, the undersigned, duly authorized by their respective Governments, have signed this convention.

Done at Paris, June 4, 1974.

Annex A

The allocation of substances to Parts I, II and III below takes account of the following criteria:

(a) persistence;

(b) toxicity or other noxious properties;

(c) tendency to bio-accumulation.

These criteria are not necessarily of equal importance for a particular substance or group of substances, and other factors, such as the location and quantities of the discharge, may need to be considered.

Part I

The following substances are included in this part:

(i) because they are not readily degradable or rendered harmless by natural processes; and

(ii) because they may either;

(a) give rise to dangerous accumulation of harmful material in the food chain, or

(b) endanger the welfare of living organisms causing undesirable changes in the marine eco-systems, or

(c) interfere seriously with the harvesting of sea foods or with other legitimate uses of the sea; and

(iii) because it is considered that pollution by these substances necessitates urgent action;

1. organohalogen compounds and substances which may form such compounds in the marine environment, excluding those which are biologically harmless, or which are rapidly converted in the sea into substances which are biologically harmless;
2. mercury and mercury compounds;
3. cadmium and cadmium compounds;
4. persistent synthetic materials which may float, remain in suspension or sink, and which may seriously interfere with any legitimate use of the sea;
5. persistent oils and hydrocarbons of petroleum origin.

Part II

The following substances are included in this part because, although exhibiting similar characteristics to the substances in Part I and requiring strict control, they seem less noxious or are more readily rendered harmless by natural processes:

1. organic compounds of phosphorus, silicon. and tin ,and substances which may form such compounds in the marine environment, excluding those which are biologically harmless, or which are rapidly converted in the sea, into substances which are biologically harmless.
2. elemental phosphorus.
3. non-persistent oils and hydrocarbons of petroleum origin.
4. The following elements and their compounds
   — arsenic,

— chromium,
— copper,
— lead,
— nickel,
— zinc.

5. Substances which have been agreed by the Commission as having a deleterious effect on the taste and/or smell of products derived from the marine environment for human consumption.

Part III

The following substances are included, in this part because, although they display characteristics similar to those of substances listed in Part I and should be subject to stringent controls with the aim of preventing and, as appropriate, eliminating the pollution which they cause, they are already the subject of research, recommendations and, in some cases, measures under the auspices of several international organizations and institutions; those substances are subject to the provisions of Article 5:

— radioactive substances, including wastes.

Annex B

Article 1
Unless the parties to the dispute decide otherwise, the arbitration procedure shall be in accordance with the provisions of this Annex.

Article 2
At the request addressed by one Contracting Party to another Contracting Party in accordance with Article 21 of the convention, an arbitral tribunal shall be constituted: The request for arbitration shall state the subject matter of the application including in particular the Articles of the convention, the interpretation or application of which is in dispute.

The claimant shall inform the Commission that he has requested the setting up of an arbitral I tribunal, stating the name of the other party of the dispute and the Articles of the convention the interpretation or application of which is in his opinion in dispute. The Commission shall forward the information thus received to all Contracting Parties to the convention.

Article 3

The arbitral tribunal shall consist of three members; each of the parties to the dispute shall appoint an arbitrator; the two arbitrators so appointed shall designate by common agreement the third arbitrator who shall be the chairman of the tribunal. The latter shall not be a national of one of the parties to the dispute, nor have his usual place of residence in the territory of one of these parties, nor be employed by any of them, nor have dealt with the case in any other capacity.

Article 4

1. If the chairman of the arbitral tribunal has not been designated within two months of the appointment of the second arbitrator, the Secretary-General of the United Nations shall, at the request of either party, designate him within a further two months' period..
2. If one of the parties to the dispute does not appoint an arbitrator within two months of receipt of the request, the other party may inform the Secretary-General of the United Nations who shall designate the chairman of the arbitral tribunal within a further two months' period. Upon designation, the chairman of the arbitral tribunal shall request the party which has not appointed an arbitrator to do so within two months. After such period, he shall inform the Secretary-General of the United Nations who shall make this appointment within a further two months' period.

Article 5

1. The arbitral tribunal shall decide according to the rules of international law and, in particular, those of this convention.
2. Any arbitral tribunal constituted under the provisions of this Annex shall draw up its own rules of procedure.

Article 6

1. The decisions of the arbitral tribunal, both on procedure and on substances, shall be taken by majority voting of its members.
2. The tribunal may take all appropriate measures in order to establish the facts. It may, at the request of one of the parties, recommend essential interim measures of protection.
3. If two or more arbitral tribunals constituted under the provisions of this Annex are seized of requests with identical or similar subjects, they may inform themselves of the procedures for establishing the facts and take them into account as far as possible.
4. The parties to the dispute shall provide all facilities necessary for the effective conduct of the proceedings.
5. The absence or default of a party to the dispute shall not constitute an impediment to the proceedings.

Article 7

1. The award of the arbitral tribunal shall be accompanied by a statement of reasons. It shall be final and binding upon the parties to the dispute.
2. Any dispute which may arise between the parties concerning the interpretation or execution of the award may be submitted by either party to the Arbitral tribunal which made the award or, if the latter cannot be seized thereof, to another arbitral

tribunal constituted for this purpose in the same manner as the first.

Article 8

The European Economic Community, like any Contracting Party to the present convention, has the right to appear as applicant or respondent before the arbitral tribunal.

Protocol, Amending the Paris Convention for the Prevention of Marine Pollution from, Land-Based Sources

The Contracting Parties to the Convention for the prevention of marine pollution from land-based sources, done at Paris on 4 June 1974 (hereinafter referred to as 'the convention);

Recalling Article 1 of the Convention, in which the Contracting Parties pledge themselves to take all possible steps to prevent pollution of the sea;

Recognizing that the. Convention does not contain provisions referring to the prevention of pollution of the maritime area through the atmosphere;

Desiring to extend the scope of the Convention to such pollution;

Have Agreed as Follows:

Article 1

The following is inserted in Article 3 of the Convention after in of subparagraph c: by emissions into the atmosphere from land or from man-made structures as defined in subparagraph in above'.

Article II

The first sentence of Article 4, paragraph 3, is amended by inserting 'and emissions into the atmosphere', after 'discharges into watercourses'.

Article III

The following is inserted at the beginning of Article 16 d of the Convention: to examine the feasibility of and, as appropriate.'

Article IV

1. This Protocol shall be open for signature at Paris from 26 March 1986 until 30 June 1986 by the States which are parties to the Convention on the date of the opening for signature of this Protocol, and by the European Economic Community.
2. This Protocol shall be subject to ratification, acceptance or approval.

Article V

After 30 June 1986 this Protocol shall be open for accession by any State referred to in Article 24 of the Convention and by the European Economic Community.

Article VI

1. This Protocol shall enter into force on the first day of the second month following the date on which the last of the contracting parties referred to in Article IV of this Protocol has deposited its instrument of ratification, acceptance, approval or accession.
2. For any other State becoming party to this protocol after its entry into force, this Protocol shall enter into force on the first day of the second month following the date on which that State has deposited its instrument of accession.
3. Any State which becomes a contracting party to this Protocol without being a contracting party to the Convention shall be considered as a contracting party to the Convention as amended by this Protocol as of the date of entry into force of this Protocol for, that State.
4. Any State which becomes a contracting party to the Convention after the entry into force of this protocol

shall be considered as a contracting party to the Convention as amended by this protocol.

5. The instruments of ratification, acceptance, approval or accession shall be deposited with the Government of the French Republic.

Article VII

The Depositary Government shall inform the Contracting Parties and those States referred to in Article 22 of the Convention of signature of this Protocol, of the deposit of instruments of ratification, acceptance, approval or accession, made pursuant to Article IV, V and VI, and of the date of entry into force of this Protocol.

Article VIII

The original of this Protocol, of which the English and French texts shall be equally authentic, shall be deposited with the Government of the French Republic.

6 Barcelona Convention for the Protection of the Mediterranean Sea against Pollution (1976)

The Contracting Parties

Conscious of the economic, social, health and cultural value of the marine environment of the Mediterranean Sea Area.

Fully aware of their responsibility to preserve this common heritage, for the benefit and enjoyment of present and future generations,

Recognizing the threat posed by pollution to -the marine environment, its ecological equilibrium, resources and legitimate uses,

Mindful of the special hydrographic and ecological characteristics of the Mediterranean Sea Area and its particular vulnerability to pollution,

Noting that existing international conventions on the subject do not cover, in spite of the progress achieved all aspects and sources of marine pollution and do not entirely meet the special requirements of the Mediterranean Sea Area,

Realizing fully the need for close co-operation among the States and international organizations concerned in a co-ordinated and comprehensive regional approach for the protection and enhancement of the marine environment in the Mediterranean Sea Area,

Have Agreed as Follows:

Article 1

Geographical coverage

1. For the purposes of this Convention, the Mediterranean Sea Area shall mean the maritime waters of the Mediterranean Sea Proper, including its gulfs and seas bounded to the west by the meridian passing through Cape Spartel lighthouse, at the entrance of the Straits of Gibraltar, and to the East by the southern limits of the Staits of the Dardanelles between Mehmetcik and Kumkale lighthouses.
2. Except as maybe otherwise provided in any protocol to this Convention the Mediterranean Sea Area shall not include internal waters of the Contracting Parties.

Article 2

Definitions

For the purposes of this Convention:

(a) "Pollution" means the introduction by man, directly or indirectly, of substances or energy into the marine environment resulting in such deleterious effects as harm to living resources, hazards to human health, hindrance to marine activities including fishing, impairment of quality for use of sea water and reduction of amenities.

(b) "organizations" means the body designated as responsible for carrying out secretariat functions pursuant to article 13 of this Convention.

Article 3

General provisions

1. The Contracting Parties may enter into bilateral or multilateral agreements, including regional or subregional agreements, for the protection of the marine environment of the mediterranean Sea against pollution, provided that such agreements are consistent with this Convention and conform to international law. Copies of such agreements between Contracting Parties to this Convention shall be communicated to the Organization.
2. Nothing in this Convention shall prejudice the codification and, development of the Law of the Sea by the United Nations Conference on the Law of the Sea convened pursuant to resolution 2750 C (XXV) of the General Assembly of the United Nations, nor the present or future claims and legal views of any State concerning the law of the sea and the nature and extent of coastal and flag State jurisdiction.

Article 4

General undertakings

1. The Contracting Parties shall individually or jointly take all appropriate measures in accordance with the provisions of this Convention and those protocols in force to which they are party, to prevent abate and combat pollution of the Mediterranean Sea Area and to protect and enhance the marine environment in that-Area.
2. The Contracting Parties shall co-operate in the formulation and adoption of protocols, in addition to the protocols opened for signature at the same time as this Convention, prescribing agreed measures, procedures and standards for the implementation of this Convention.
3. The Contracting Parties further pledge themselves to

promote, within the international bodies considered to be competent by the Contracting Parties, measures concerning the protection of the marine environment in the Mediterranean Sea Area from all types and sources of pollution.

Article 5

Pollution caused by dumping from ships and aircraft The Contracting Parties shall take all appropriate measures to prevent and abate pollution of the Mediterranean Sea Area caused by dumping from ships and aircraft.

Article 6

Pollution from ships

The Contracting Parties shall take all measures in conformity with international law to prevent, abate, and combat pollution of the Mediterranean Sea Area caused by discharges from ships and to ensure the effective implementation in that Area of the rules which are generally recognized at the international level relating to the control of this type of pollution.

Article 7

Pollution resulting from exploration and exploitation of the continental shelf and the seabed and its subsoil

The *Contracting Parties* shall take all appropriate measures to prevent, abate and combat pollution of the Mediterranean Sea Area resulting from exploration and exploitation of the continental shelf and the seabed and its subsoil.

Article 8

Pollution from land-based sources

The Contracting Parties shall take all appropriate measures to prevent, abate and combat pollution of the Mediterranean Sea Area caused by discharges from rivers, coastal establishments or outfalls, or emanating from any other land-based sources within their territories.

Article 9

Co-operation in dealing with pollution emergencies

1. The Contracting Parties shall co-operate in taking the necessary measures for dealing with pollution emergencies in the, Mediterranean Sea Area, whatever the causes of such emergencies, and reducing or eliminating damage resulting therefrom.
2. Any Contracting Party which becomes aware of any pollution emergency in the Mediterranean Sea Area shall without delay notify the Organization and, either through the Organization or directly, any Contracting Party likely to be affected by such emergency.

Article 10

Monitoring

1. The Contracting Parties shall endeavour to establish, in close co-operation with the international bodies which they consider competent, complementary or joint programmes including, as appropriate programmes at the bilateral or multi-lateral levels, for pollution monitoring in the Mediterranean Sea Area and shall endeavour to establish a pollution-monitoring system for that Area.
2. For this purpose, the Contracting Parties shall designate the competent authorities responsible for pollution monitoring within areas under their national jurisdiction and participate as far as practicable in international arrangements for pollution monitoring in areas beyond national jurisdiction.
3. The Contracting Parties undertake to co-operate in the formulation, adoption and implementation of such annexes to this Convention as maybe required to prescribe common procedures and standards for pollution monitoring.

Article 11

Scientific and technological co-operation

1. The Contracting Parties undertake as far as possible to co-operate directly, or when appropriate through competent regional or other international organizations, in the fields of science and technology, and to exchange data as well as other scientific information for the purpose of this Convention.
2. The Contracting Parties undertake as far as possible to develop and co-ordinate their national research programmes relating to all types of marine pollution in the Mediterranean Sea Area and to co-operate in the establishment and implementation of regional and other international research programmes for the purposes of this Convention.
3. The Contracting Parties undertake to co-operate in the provision of technical and other possible assistance in fields relating to marine pollution, with priority to be given to the special needs of developing countries in the Mediterranean region.

Article 12

Liability and compensation

The Contracting Parties undertake to co-operate as soon as possible in the formulation and adoption of appropriate procedures for the determination of liability and compensation for damage resulting from the pollution of the marine environment deriving from violations of the provisions of this Convention and applicable protocols.

Article 13

Institutional arrangements

The Contracting Parties designate the United Nations

Environment Programme as responsible for carrying out the following secretariat functions:

- to convene and prepare the meetings of Contracting Parties and conferences provided for in articles 14, 15, and 16;
- to transmit to the Contracting Parties notifications, reports and other information received in accordance with articles 3, 9 and 20;
- to consider inquiries by, and information from, the Contracting Parties, and to consult with them on questions relating to this Convention and the protocols and annexes thereto;
- to perform the functions assigned to it by the protocols to this Convention;

  to perform such other functions as may be assigned to it by the Contracting Parties;
- to ensure the necessary co-ordination with other international bodies which the Contracting Parties consider competent, and in particular to enter into such administrative arrangements as may be required for the effective discharge of the secretariat functions.

Article 14

Meetings of the Contracting Parties

1. The Contracting Parties shall hold ordinary .Meetings once every two years, and extraordinary meetings at any other time deemed necessary, upon the request of the Organization or at the request of any Contracting Party, provided that such requests are supported by at least two Contracting Parties.
2. It shall be the function of the meetings of the Contracting Parties to keep under review the implementation of this Convention and the protocols and, in particular:

(i) to review generally the inventories carried out, by Contracting Parties and competent international organizations on the state of marine pollution and its effects in the Mediterranean Sea Area;

(ii) to consider reports submitted by the Contracting Parties under article 20;

(iii) to adopt, review and amend as required the annexes to this Convention and to the protocols, in accordance with the procedure established in article 17;

(iv) to make recommendations regarding the adoption of any additional protocols or any amendments to this Convention or the protocols in accordance with the provisions of articles 15 and 16;

(v) to establish working groups as required to consider any matters related to this Convention and the protocols and annexes;

(vi) to consider and undertake any additional action that may be required for the achievement of the purposes of this Convention and the protocols.

Article 15

Adoption of additional protocols

1. The Contracting Parties, at a diplomatic conference, may adopt additional protocols to this Convention pursuant to paragraph 2 of article 4.
2. A diplomatic conference for the purpose of adopting additional protocols shall be convened by the Organization at the request of two thirds of the Contracting Parties.
3. Pending the entry into force of this Convention the Organization may, after consulting with the

signatories to this Convention, convene a diplomatic conference for the purpose of adopting protocols.

Article 16

Amendment of the Convention or Protocols

1. Any Contracting Party to this Convention may propose amendments to the Convention. Amendments shall be adopted by a diplomatic conference which shall be convened by the Organization at the request to two-thirds of the Contracting Parties.
2. Any Contracting Party to this Convention may propose amendments to any protocol. Such amendments shall be adopted by a diplomatic conference which shall be convened by the Organization at the request to two-thirds of the Contracting Parties to the protocol concerned.
3. Amendments to this Convention shall be adopted by a three-fourths majority vote of the Contracting Parties to the Convention which are represented at the diplomatic conference, and shall be submitted by the Depositary for acceptance by all Contracting Parties to the Convention. Amendments to any protocol shall be adopted by a three-fourths majority vote of the Contracting Parties to such protocol which are represented at the diplomatic conference, and shall be submitted by the Depositary for acceptance by all Contracting Parties to such protocol.
4. Acceptance of amendments shall be notified to the Depositary in writing. Amendments adopted in accordance with paragraph 3 of this article shall enter into force between I Contracting Parties having accepted such amendments on the thirtieth day following the receipt by the Depositary of notification of their acceptance by at least three-

fourths, of the Contracting Parties to this Convention or to the protocol concerned, as the case may be.

5. After the entry into force of an amendment of this Convention or to a protocol, any new Contracting Party to this Convention or such protocol shall become a Contracting Party to the instrument as amended.

Article 17

Annexes and amendments to Annexes

1. Annexes to this Convention or to any protocol shall form an integral part of the Convention or such protocol, as the case may be.
2. Except as may be otherwise provided in any protocol, the following procedure shall apply to the adoption and entry into force of any amendments of annexes to this Convention or to any Protocol, with the exception of amendments to the Annex on arbitration:
   (i) any Contracting Party may propose amendments to the annexes to this Convention or to protocols at the meetings referred to in article 14;
   (ii) such amendments shall be adopted by a three-fourths majority vote of the Contracting Parties to the instrument in question;
   (iii) the Depositary shall without delay communicate the amendments-so adopted to all Contracting Parties;
   (iv) any Contracting Party that is unable to approve an amendment to the annexes to this Convention or to any protocol shall so notify in writing the Depositary within a period determined by the Contracting Parties concerned when adopting the amendment;

(v) the Depositary shall without delay notify all Contracting Parties of any notification received pursuant to the preceding sub-paragraph;

(vi) on expiry of the period referred to in sub paragraph (iv) above, the amendment to ' the annex shall become effective for all Contracting Parties to this Convention or to the protocol concerned which have not submitted a notification in accordance with the provisions of that sub-paragraph.

3. The adoption and entry into force of a new annex to this Convention or to any protocol shall be subject to the same procedure as for the adoption and entry into force provided that, if any amendment to the Convention or the protocol concerned is involved, the new annex shall not enter into force until such time as the amendment to the Convention or the protocol concerned enters into force.

4. Amendments to the annex on arbitration shall be considered to be amendments to this Convention and shall be proposed and adopted in accordance with the procedures set out in article 16 above.

Article 18

Rules of procedure and, financial rules

1. The Contracting Parties shall adopt rules of procedure for their meetings and conferences envisaged in articles 14, 15, and 16 above.

2. The Contracting Parties shall adopt financial rules, prepared in consultation with the Organization, to determine, in particular, their financial participation.

Article 19

Special exercise of voting right

Within the areas of their competence, the European Economic Community and any regional economic grouping referred to in article 24 of this Convention shall exercise their right to vote with a number of votes equal to the number of their member States which are Contracting Parties to this Convention and to one or more Protocols; the European Economic Community and any grouping as referred to above shall not exercise their right to vote in cases where the member States concerned exercise theirs, and conversely.

Article 20

Reports

The Contracting Parties shall transmit to the Organization reports on the measures adopted in implementation of this Convention and of Protocols to which they are Parties, in such form and at such intervals as the meetings of Contracting Parties may determine.

Article 21

Compliance control

The Contracting Parties undertake to co-operate in the development of procedures enabling them to Control the application of this Convention and the Protocols.

Article 22

Settlement of disputes

1. In case of a dispute between Contracting Parties as to the interpretation or application of this Convention or the Protocols, they shall seek a settlement of the dispute through negotiation or any other peaceful means of their own choice.
2. If the parties concerned cannot settle their dispute through the means mentioned in the preceding paragraph, the dispute shall upon common agreement be submitted to arbitration under the conditions laid down in Annex A to this Convention.

3. Nevertheless, the Contracting Parties may at any time declare that they recognize as compulsory *ipso facto* and without special agreement, in relation to any other Party accepting the same obligation, the application of the arbitration procedure in conformity with the provisions of Annex A. Such declaration shall be notified in writing to the Depositary, who shall communicate it to the other Parties.

Article 23

Relationship between the Convention and protocols

1. No one may become a Contracting Party to this Convention unless it becomes at the same time a Contracting Party to at least one of the protocols. No one may become a Contracting Party to a protocol unless it is, or becomes at the same time a Contracting Party to this Convention.
2. Any protocol to this Convention shall be binding only on the Contracting Parties to the protocol in question.
3. Decisions concerning any protocol pursuant to articles 14, 16 and 17 of this Convention shall be taken only by the parties to the protocol concerned.

Article 24

Signature

This Convention, the Protocol for the Prevention of Pollution of the Mediterranean Sea by Dumping from Ships and Aircraft and the Protocol concerning Cooperation in Combating Pollution of the Mediterranean Sea by Oil and Other Harmful Substances in Cases of Emergency shall be open for signature in Barcelona on 16 February 1976 and in Madrid from 17 February 1976 to 16 February 1977 by any State invited as a participant in the Conference of Plenipotentiaries of the

Coastal States of the Mediterranean Region on the Protection of the Mediterranean Sea, held in Barcelona from 2 to 16 February 1976, and by any State entitled to sign any protocol in accordance with the provisions of such Protocol. They shall also be open until the same date for signature by the European Economic Community and by any similar regional economic grouping at least one member of which is a coastal State of the Mediterranean Sea Area and which exercises competencies in fields covered by this Convention, as well as by any protocol affecting them.

Article 25

Ratification, acceptance or approval

This Convention and any protocol thereto shall be subject to ratification, acceptance, or approval. Instruments or ratification, or approval shall be deposited with the Government of Spain, which will assume the function of Depositary.

Article 26

Accession

1. As from 17 February 1977, the present Convention-, the Protocol for the Prevention of Pollution of the Mediterranean Sea by Dumping from ships and Aircraft, and the protocol concerning co-operation in. Combating Pollution of the Mediterranean Sea by Oil and other Harmful Substances in Cases of Emergency shall be open for accession by the States, by the European Economic Community and by any grouping as referred to in article 24.
2. After the entry into force of the Convention and of any protocol, any State not referred to in article 24 may accede to this Convention and to any protocol, subject to prior approval by three-fourths of the Contracting Parties to the protocol concerned.

3. Instruments of accession shall be deposited with the Depositary.

Article 27

Entry into force

1. This Convention shall enter into force on the same date as the protocol first entering into force.
2. The Convention shall also enter into force with regard to the States, the European Economic Community and any regional economic grouping referred to in article 24 if they have complied with the formal requirements ,for becoming Contracting Parties to any other protocol not yet entered into force.
3. Any protocol to this Convention, except as otherwise provided in such protocol, shall enter into force on the thirtieth day following the date of deposit of at least six instruments of ratification, acceptance, or approval of, or accession to such protocol by the Parties referred to in article 24.
4. Thereafter, this Convention and any protocol shall enter into force with respect to any State, the European Economic Community and any regional economic grouping' referred to in article 24 on the thirtieth day following the date of deposit of the instruments of ratification acceptance approval or accession.

Article 28

Withdrawal

1. At any time after three years from the date of entry into force of this Convention, any Contracting Party may withdraw from this Convention by giving written notification of withdrawal.
2. Except as may be otherwise provided in any protocol to this Convention any Contracting Party

may, at any time after three years from the date of entry into force of such protocol, withdraw from such protocol by giving written notification of withdrawal.

3. Withdrawal shall take effect 90 days after the date on which notification of withdrawal Is received by the Depositary.
4. Any Contracting Party which withdraws from this Convention shall be considered as also having withdrawn from any protocol to which it was a Party.
5. Any Contracting Party which, upon its withdrawal from a protocol, is no longer a Party to any protocol to this Convention, shall be considered as also having withdrawn from this Convention.

Article 29

Responsibilities of the Depositary

1. The Depositary shall inform the Contracting Parties, any other Party referred to in article. 24, and the Organizations:
   (i) of the signature of this Convention and of any protocol thereto, and of the deposit of instruments of ratification, acceptance, approval or accession in accordance with articles 24, 25 and 26;
   (ii) of the date on which the Convention and any protocol with come into force in accordance with the provisions of article 27;
   (iii) of notifications of withdrawal made in accordance with article 28;
   (iv) of the amendments adopted with respect to the Convention and to any protocol, their acceptance by the Contracting Parties and the date of entry into force of those amendments in accordance with the provisions of article 16;

(v) of the adoption of new annexes and of the amendment of any annex in accordance with article 17;

(vi) of declarations recognizing as compulsory the application of the arbitration procedure mentioned in paragraph 3 of article 22.

2. The original of this Convention and of any protocol thereto shall be deposited with the Depositary, the Government of Spain, which shall send certified copies thereof to the Contracting Parties, to the Organization, and to the Secretary-General of the United Nations for registration and publication in accordance with Article 102 of the United Nations Charter.

In Witness Whereof the undersigned, being duly authorized by their respective Governments, have signed this Convention

Done at Barcelona on 16 February 1976 in a single copy in the Arabic, English, French and Spanish languages, the four texts being equally authoritative.

Annex A

Arbitration

Article 1

Unless the Parties to the dispute otherwise agree, the arbitration procedure shall be conducted in accordance with the provisions of this Annex.

Article 2

1. At the request addressed by one Contracting Party to another Contracting Party in accordance with the provisions of paragraph 2 or paragraph 3 of article 22 of the Convention, an arbitral tribunal shall be constituted. The request for arbitration shall state the subject matter of the application including, in particular, the articles of the Convention or the

Protocols, the interpretation or application of which is in dispute.

2. The claimant party shall inform the Organization that it has requested the setting up of an arbitral tribunal, stating the name of the other party to the dispute and articles of the Convention or the Protocols the interpretation or application of which is in its opinion in dispute. The organization shall forward the information thus received to all Contracting Parties to the Convention.

Article 3

The arbitral tribunal shall consist of three members: each of the parties to the dispute shall appoint an arbitrator; the two arbitrators so appointed shall designate by common agreement the third arbitrator who shall be the chairman of the tribunal. The latter shall not be a national of one of the parties to the dispute, nor have his usual place of residence in the territory of one of these parties, nor be employed by any of them, nor have dealt with the case in any other capacity.

Article 4

1. If the chairman of the arbitral tribunal has not been designated within two months of the appointment of the second arbitrator, the Secretary-General of the United Nations shall, at the request of the most diligent party designate him within a further two months period.

2. If one of the parties of the dispute does not appoint an arbitrator within two months of receipt of the request, the other party may inform the Secretary-General of the United Nations who shall designate the chairman of the arbitral tribunal within a further two months' period. Upon designation, the chairman of the arbitral tribunal shall request the party which has not appointed an

arbitrator to do so within two months. After such period, he shall inform the Secretary-General of the United Nations, who shall make this appointment within a further two months' period.

Article 5

1. The arbitral tribunal shall decide according to the rules of international law and, in particular, those of this Convention and the protocols concerned.
2. Any arbitral tribunal constituted under the provisions of this Annex shall draw up its own rules of procedure.

Article 6

1. The decisions of the arbitral tribunal, both on procedure and on substance, shall be taken by majority vote of its members.
2. The tribunal may take all appropriate measures in order the establish the facts. It may, at the request of one of the parties, recommend essential interim, measures of protection.
3. If two or more arbitral tribunals constituted under the provisions of this Annex are seized of requests with identical or similar subjects, they may inform themselves of the procedures for establishing the facts and take them into account as far as possible.
4. The parties to the dispute shall provide all facilities necessary for the effective conduct of the proceedings.
5. The absence or default of a party to the dispute shall not constitute an impediment to the proceedings.

Article 7

1. The award of the arbitral tribunal shall be accompanied by a statement of reasons. It shall be final and binding upon the parties to the dispute.
2. Any dispute which may arise between the parties concerning the interpretation or execution of the

award may be submitted by the most diligent party to the arbitral tribunal which made the award or, if the latter cannot be seized thereof, to another arbitral tribunal constituted for this purpose in the same manner as the first.

Article 8

The European Economic Community and any regional economic grouping referred to in article 24 of the Convention, like any Contracting Party to the Convention, are empowered to appear as complainants or as respondents before the arbitral tribunal.

Protocol for the Prevention of Pollution of the Mediterranean Sea by Dumping from Ships and Aircraft

The Contracting Parties to the Present Protocol,

Being Parties to the Convention for the Protection of the Mediterranean Sea against Pollution,

Recognizing the danger posed to the marine environment by pollution caused by the dumping of wastes or other matter from ships and aircraft,

Considering that the coastal States of the Mediterranean Sea have a common interest in protecting the marine environment from this danger,

Bearing in mind the Convention on the Prevention of Marine Pollution by Dumping of Wastes and other Matter, adopted in London in 1972.

Have Agreed as Follows:

Article 1

The Contracting Parties to this Protocol (hereinafter referred to as "the Parties") shall take all appropriate measures to prevent and abate pollution of the Mediterranean Sea Area caused by dumping from ships and aircraft.

Article 2

The area to which this Protocol applies shall be the Mediterranean Sea Area as defined in article 1 of the Convention for the Protection of the Mediterranean Sea against Pollution (hereinafter referred to as "the Convention").

Article 3

For the purposes of this Protocol:

1. "ships and aircraft" means waterborne or airborne craft of any type whatsoever. This expression includes air-cushioned craft and floating craft whether self-propelled or not, and platforms and other man-made structures at sea and their equipment.
2. "Wastes or other matter" means material and substances of any kind, form or description.
3. 'Dumping" means:
   (a) any deliberate disposal at sea of wastes or other matter from ships or aircraft;
   (b) any deliberate disposal at sea of ships or aircraft.
4. "Dumping" does not include:
   (a) the disposal at sea of wastes or other matter incidental to, or derived from the normal operations of vessels, or aircraft and their equipment, other than wastes or other matter transported by or to vessels, or aircraft, operating for the purpose of disposal of such matter or derived from the treatment of such wastes or other matter on such vessels or aircraft;
   (b) placement of matter for a purpose other than the mere disposal thereof, provided that such placement is not contrary to the aims of this Protocol.

5. "Organization" means the body referred to in article 13 of the Convention.

Article 4

The dumping into the Mediterranean Sea Area of wastes or other matter listed in Annex I to this Protocol is prohibited.

Article 5

The dumping into the Mediterranean Sea Area of wastes or other matter listed in Annex II to this Protocol requires, in each case, a prior special permit from the competent national authorities.

Article 6

The dumping into the Mediterranean Sea Area of all other wastes or other matter requires a prior general permit from the competent national authorities.

Article 7

The permits referred to in articles 5 and 6 above shall be issued only after careful consideration of all the factors set forth in Annex III to this Protocol. The Organization shall receive records of such permits.

Article 8

The provisions of articles 4, 5 and 6 shall not apply in case of force majeure due to stress of weather or any other cause when human life or the safety of a ship or aircraft is threatened. Such dumpings shall immediately be reported to the Organization, either through the Organization or directly to any Party or Parties likely to be affected, together with full details of the circumstances and of the nature and quantities of the wastes or other matter dumped.

Article 9

If a Party in a critical situation of an exceptional nature considers that wastes or other matter listed in Annex I to this Protocol cannot be disposed of on land without unacceptable danger or damage, above all for the safety of human life, the Party concerned shall forthwith consult the Organization The Organization, after consulting the Parties to this Protocol, shall recommend methods of storage or the most satisfactory means of destruction or disposal under the prevailing circumstances. The Party shall inform the Organization of the steps adopted in pursuance of these recommendations. The Parties pledge themselves to assist one another in such situations.

Article 10

1. Each Party shall designate one or more competent authorities to:
   (a) issue the special permits provided for in article 5;
   (b) issue the general permits provided for in article 6;
   (c) keep records of the nature and quantities of the wastes or other matter permitted to be dumped and the location, date and method of dumping.
2. The competent authorities of each Party shall issue the permits provided for in articles 5 and 6 in respect of the wastes or other matter intended for dumping to be:
   (a) loaded in its territory;
   (b) loaded by a ship or aircraft registered in its territory or flying its flag, when the loading occurs in the territory of a State not Party to this Protocol.

Article 11

1. Each Party shall apply the measures required to implement this Protocol to all:
   (a) ships and aircraft registered in its territory or flying its flag;

(b) ships and aircraft loading in its territory wastes or other matter which are to be dumped;

(c) ships and aircraft believed to be engaged in dumping in areas under its jurisdiction in this matter.

2. This Protocol shall not apply to any ships or aircraft owned or operated by a State Party to this Protocol and used for the time being only on Government noncommercial service. However, each Party shall ensure by the adoption of appropriate measures not impairing the operations or operational capabilities of such ships or aircraft owned or operated by it, that such ships and aircraft act in a manner consistent, so far as is reasonable and practicable, with this Protocol.

Article 12

Each Party undertakes to issue instructions to its maritime inspection ships and aircraft and to either appropriate services to report to its authorities any incidents or conditions in the Mediterranean Sea Area which gives rise to suspicions that dumping in contravention of the provisions of this Protocol has occurred or is about to occur. The Party shall, if it considers it appropriate, report accordingly to any other Party concerned.

Article 13

Nothing in this Protocol shall affect the right of each Party to adopt other measures, in accordance with international law, to prevent pollution due to dumping.

Article 14

1. Ordinary meetings of the Parties to this Protocol shall be held in conjunction with ordinary meetings of the Contracting Parties to the Convention held pursuant to article 14 of the Convention. The Parties to this Protocol may also hold extraordinary

meetings in conformity with article 14 of the Convention.

2. It shall be the function of the meetings of the Parties to this Protocol:
   (a) to keep under review the implementation of this Protocol, and to consider the efficacy of the measures adopted and the need for any other measures, in particular in the form of annexes;
   (b) to study and consider the records of the permits issued in accordance without articles 5, 6 and 7, and of the dumping which has taken place;
   (c) to review and amend as required any Annex to this Protocol;
   (d) to discharge such other functions as may be appropriate for the implementation of this Protocol.
3. The adoption of amendments to the Annexes to this Protocol pursuant to article 17 of the Convention shall require a three-fourths majority of the Parties.

Article 15

1. The provisions of the Convention relating to any Protocol shall apply with respect to the present Protocol.
2. The rules of procedure and the financial rules adopted pursuant to article 18 of the Convention shall apply with respect to this Protocol, unless the Parties to this Protocol agree otherwise.

In Witness whereof the undersigned, being duly authorized by their respective Governments, have signed this Protocol.

Done at Barcelona on 16 February 1976 in a single copy in the Arabic, English, French and Spanish languages; the four texts being equally authoritative.

Annex I

A. The following substances and materials are listed for the purpose of article 4 of the Protocol.

1. Organohalogen compounds and compounds which may form substances in the marine environment, excluding those which are non-toxic or which are rapidly converted in the sea into substances which are biologically harmless, provided that they do not made edible marine organisms unpalatable.
2. Organosilicon compounds and compounds which may form such substances in the marine environment excluding those which are non-toxic or which are rapidly converted in the sea into substances which are biologically harmless, provided that they do not make edible marine organisms unpalatable.
3. Mercury and mercury compounds.
4. Cadmium and cadmium compounds.
5. Persistent plastic and other persistent synthetic materials which may materially interfere with fishing or navigation, reduce amenities, or interfere with other legitimate uses of the sea.
6. Crude oil and hydrocarbons which may be derived from petroleum, and any mixtures containing any of these, taken on board for the purpose of dumping.
7. High-and medium- and low-level radioactive wastes or other high-and medium-and low-level radioactive matter to be defined by the International Atomic Energy Agency.
8. Acid and alkaline compounds of such composition and in such quantity that they may seriously impair the quality of sea water. The composition and quantity to be taken into consideration shall be determined by the Parties in accordance with the

procedure laid down in article 14, paragraph 3, of this Protocol.

9. Materials in whatever form (e.g. solids, liquids, semiliquids, gases, or in a living state) produced for biological and chemical warfare, other than those rapidly rendered harmless by physical, chemical or biological processes in the sea provided that they do not:

   (i) make edible marine organisms unpalatable; or

   (ii) endanger human or animal health.

B. This Annex does not apply to wastes or other materials, such as sewage sludge and dredge spoils, containing the substances referred to in paragraphs 1-6 above as trace contaminants. The dumping of such wastes shall be subject to the provisions of Annexes II and III as appropriate.

Annex II

The following wastes and other matter the dumping of which requires special care are listed for the purposes of article 5.

1. (i) arsenic, lead, copper, zinc, beryllium, chromium, nickel, vanadium, selenium, antimony and their compounds;

   (ii) cyanides and fluorides;

   (iii) pesticides and their by-products not covered in Annex I;

   (iv) syntetic organic chemicals, other than those referred to in Annex I, likely to produce harmful effects on marine organisms or to make edible marine organisms unpalatable.

2. (i) acid and alkaline compounds the composition and quantity of which have not yet been determined in accordance with the procedure referred to in Annex I, paragraph A.8.

(ii) acid and alkaline compounds not covered by Annex 1, excluding compounds to be dumped in quantities below thresholds which shall be determined by the Parties in accordance with the procedure laid down in article 14, paragraph 3 of this Protocol.

3. containers, scrap metal and other bulky wastes liable to sink to the sea bottom which may present a serious obstacle to fishing or navigation.
4. substances which, though of a non-toxic nature may become harmful owing, to the quantities in which they are dumped, or which are liable to reduce amenities seriously or to endanger human life or marine organisms or to interfere with navigation.
5. radioactive waste or other radioactive matter which will not be included in Annex I. In the issue of permits for the dumping of this matter, the Parties should take full account of the recommendations of the competent international body in this field, at present the international Atomic Energy Agency.

Annex III

The factors to be considered in establishing criteria governing the issue of permits for the dumping of matter at sea taking into account article 7 include:

A. Characteristics and composition of the matter

1. Total amount and average compositions of matter dumped (e.g. per year).
2. Form (e.g. solid, sludge, liquid or gaseous).
3. Properties: physical (e.g. solubility and density), chemical and bio-chemical (e.g. oxygen demand, nutrients) and biological (e.g. presence of viruses, bacteria, yeasts, parasites).
4. Toxicity.

5. Persistence: physical, chemical and biological.
6. Accumulation and biotransformation in biological materials or sediments.
7. Susceptibility to physical, chemical and biochemical changes and interaction in the aquatic-environment with other dissolved organic and inorganic materials.
8. Probability of production of taints or other changes reducing marketability of resources (fish, shellfish, etc.).

B. Characteristics of dumping site and method of deposit

1. Location (e.g. co-ordinates of the dumping area, depth and distance from the coast), location in relation to other areas (e.g. amenity areas, spawning, nursery and fishing areas and exploitable resources).
2. Rate of disposal per specific period (e.g. quantity per day, per week, per month).
3. Methods of packaging and containment, if any.
4. Initial dilution achieved by proposed method of release, particularly the speed of the ship.
5. Dispersal characteristics (e.g. effects of currents, tides and wind on horizontal transport and vertical mixing).
6. Water characteristics (e.g. temperature, pH, salinity, stratification, oxygen indices of pollution-dissolved oxygen (DO), chemical oxygen demand (CDO), biochemical oxygen demand (BOD), nitrogen present in organic and mineral form, including ammonia, suspended matter, other nutrients and productivity).
7. Bottom characteristics (e.g. topography, geochemical and geological characteristics and biological productivity).

8. Existence and effects of other dumpings which have been made in the dumping area (e.g. heavy metal background reading and organic carbon content).
9. When issuing a permit for dumping, the Contracting Parties shall endeavour to determine whether an adequate scientific basis exists for assessing the consequences of such dumping in the area concerned, in accordance with foregoing provisions and taking into account seasonal variations.

C. General considerations and conditions

1. Possible effects on amenities (e.g. presence of floating or stranded material, turbidity, objectionable odour, discoloration and foaming).
2. Possible effects on marine life, fish and shellfish culture, fish stocks and fisheries, seaweed harvesting and culture.
3. Possible effects on other uses of the sea (e.g. impairment of water quality for industrial use, underwater corrosion of structures, interference with ship operations from floating materials, interference with fishing or navigation through deposit of waste or solid objects on the sea floor and protection or areas of special importance for scientific or conservation purposes.
4. The practical availability of alternative land-based methods of treatment disposal or elimination, or of treatment to render the matter less harmful for sea dumping.

Protocol Concerning Co-operation in Combating Pollution of the Mediterranean Sea by Oil and other Harmful, Substances in Cases of Emergency

The Contracting Parties to the Present Protocol,

Being Parties to the Convention for the Protection of the Mediterranean Sea against Pollution.

Recognizing that grave pollution of the sea by oil and other harmful substances in the Mediterranean Sea Area involves a danger for the coastal States and the marine eco-system.

Considering that the co-operation of all the coastal States of the Mediterranean is called for to combat this pollution.

Bearing in mind the International Convention for the Prevention of Pollution from Ships, 1973, the International Convention relating to Intervention on the High Seas in Cases of Oil Pollution Casualties, 1969, as well as the Protocol relating to Intervention on the High Seas in Cases of Marine Pollution by Substances Other than Oil, 1973.

Further taking into account the International Convention on Civil Liability for Oil Pollution Damage, 1969.

Have Agreed as follows:

Article 1

The Contracting Parties to this Protocol (hereinafter referred to as "the Parties') shall co-operate in taking the necessary measures in cases of grave and imminent danger to the marine environment, the coast or related interests of one or more of the Parties due to the presence of massive quantities of oil or other harmful substances resulting from accidental causes or an accumulation of small discharge which are polluting or threatening to pollute the sea within the area defined in Article 1 of the Convention for the Protection of the Mediterranean Sea against Pollution, (hereinafter referred to as "the Convention")

Article 2

For the purpose of this Protocol, the term "related

interests" means the interests of a coastal State directly affected or threatened and concerning, among others:

(a) activities in coastal waters in ports or estuaries, including fishing activities;

(b) the historical and tourist appeal of the area in question, including water sports and recreation;

(c) the health of the coastal population;

(d) the preservation of living resources.

Article 3

The Parties shall endeavour to maintain and promote, either individually or through bilateral or multilateral cooperation, their contingency plans and means for combating pollution of the sea by oil and other harmful substances. These means shall include, in particular, equipment, ships, aircraft and manpower prepared for operations in cases of emergency.

Article 4

The Parties shall develop and apply, either individually or through bilateral or multilateral co-operation, monitoring activities covering the Mediterranean Sea Area in order to have as precise information as possible on the situations referred to in article 1 of this protocol.

Article 5

In the case of release or loss overboard of harmful substances in packages, freight containers, portable tanks or road and rail tank wagons, the Parties shall co-operate as far as practicable in the salvage and recovery of such substances so as to reduce the danger of pollution of the marine environment.

Article 6

1. Each Party undertakes to disseminate to the other Parties information concerning:

(a) the competent national organization or authorities responsible for combating pollution of the sea by oil and other harmful substances;

(b) the competent national authorities responsible, for receiving reports of -pollution of the sea by oil and other harmful substances and for dealing with matters concerning measures of assistance between Parties;

(c) new ways in which pollution of the sea by oil and other harmful substances may be avoided, new measures of combating pollution and the development of related research programmes.

2. Parties which have agreed to exchange information directly between themselves shall nevertheless communicate such information to the regional centre. The latter shall 'communicate this information to the other Parties and, on a basis of reciprocity, to coastal States of the Mediterranean Sea Area which are not Parties to this Protocol.

Article 7

The Parties undertake to co-ordinate the utilization of the means of communication at their disposal in order to ensure, with the necessary speed and reliability, the reception, transmission and dissemination of all reports and urgent information which relate to the occurrence and situations referred to in article 1. The regional centre shall have the necessary means of communication to enable it to participate in this co-ordinate efforts and, in particular, to fulfil the functions assigned to it by paragraph 2 of article 10.

Article 8

Each Party shall issue instructions to the masters of ships flying its flag to the pilots of aircraft registered in its territory requiring them to report by the most rapid and adequate channels in the circumstances, and in decordance with Annex I to this Protocol, either to a Party or to the regional centre:

(a) all accidents causing or likely to cause pollution of the sea by oil or other harmful substances;

(b) the presence, characteristics and extent of spillages of oil or other harmful substances observed at sea which are likely to present a serious and imminent threat to the marine environment or to the coast or related interests of one or more of the Parties.

2. The information collected in accordance with paragraph 1 shall be communicated to the other Parties likely to be affected by the pollution:

(a) by the Party which has received the information, either directly or preferably through the regional centre; or

(b) by the regional centre.

In case of direct communication between Parties, the regional centre shall be informed of the measures taken by the Parties.

3. In consequence of the application of the provisions of paragraph 2, the Parties are not bound by the obligation laid down in article 9, paragraph 2, of the Convention.

Article 9

1. Any Party faced with a situation of the kind defined in article 1 of this Protocol shall:

(a) make the necessary assessments of the nature and extent of the casualty or emergency or, as the case may be, of the type and approximate quantity of oil or other harmful substances and the direction and speed of drift of the spillage;

(b) take every practicable measure to avoid or reduce the effects of pollution;

(c) immediately inform all other Parties, either directly or through the regional centre, of these assessments and of any action which it has taken or which it intends to take to combat the pollution.

(d) continue to observe the situation for as long as possible and report thereon in accordance with article 8.

2. Where action is taken to combat pollution originating from a ship, all possible measures shall be taken to safeguard the persons present on board and, to the extent possible the ship itself. Any Party which takes such action shall inform the Inter-Governmental Maritime Consultative Organization.

Article 10

1. Any Party requiring assistance for combating pollution by oil or other harmful substances polluting or threatening to pollute its coasts may call for assistance from other Parties, either directly or through the regional centre referred to in article 6, starting with the Parties which appear likely to be affected by the pollution. This assistance may comprise, in particular, expert advice and the supply to or placing at the disposal of the Party concerned of products, equipment and nautical facilities. Parties so requested shall use their best endeavours to render this assistance.

2. Where the Parties engaged in an operation to combat pollution cannot agree on the organization of the operation, the regional centre may, with their approval, coordinate the activity of the facilities put into operation by these Parties.

Article 11

The application of the relevant provisions of articles 6, 7, 8, 9 and 10 of this Protocol relating to the regional centre shall be extended, as appropriate, to subregional centres in the event of their establishment, taking into account their objectives and function's and their relationship with the said regional centre.

Article 12

1. Ordinary meetings of the Parties to this Protocol shall be held in conjunction with ordinary meetings of the Contracting Parties to the Convention, held pursuant to article 14 of the Convention. The Parties to this Protocol may also hold extraordinary meetings as provided in article 14 of the Convention.
2. It shall be the function of the meetings of the Parties to this Protocol, in particular:
   (a) to keep under review the implementation of this Protocol, and to consider the efficacy of the measures adopted and the need for any other measures, in particular in the form of Annexes;
   (b) to review and amend as required any. Annex to this Protocol;
   (c) to discharge such other functions as may be appropriate for implementation of this Protocol.

Article 13

1. The provisions of the Convention relating to any Protocol shall apply with respect to the present Protocol.
2. The rules of procedure and the financial rules adopted pursuant to article 18 of Convention shall apply with respect to this Protocol, unless the Parties to this Protocol agree otherwise.

In witness whereof the undersigned, being duly authorized by their respective Governments, have signed this Protocol.

Done at Barcelona on 16 February 1976 in a single copy in the Arabic, English, French and Spanish languages, the four texts being equally authoritative.

Annex I

Contents of the report to be made pursuant to article 8 to this Protocol

1. Each report shall, as far as possible, contain, in general:
   (a) the identification of the source of pollution (identity of the ship, where appropriate);
   (b) the geographic position, time and date of the occurrence of-the incident or of the observation-,
   (c) the wind and-sea conditions prevailing in the area.
   (d) where the pollution originates from a ship, relevant details respecting conditions of the ship.

Each report shall contain, whenever possible, in particular:

   (a) a clear indication or description of the harmful substances involved, providing the correct technical names of such substances (trade names should not be used in place of the correct technical names);
   (b) a statement or estimate of the quantities, concentrations and likely conditions of harmful substances discharged or likely to be discharged into the sea;
   (c) where relevant, a description of the packaging and identifying marks; and
   (d) the name of the consignor, consignee or manufacturer.

3. Each report shall clearly indicate, whenever possible, whether the harmful substance discharged or likely to be discharged is oil or a noxious liquid, solid or gaseous substance and whether such substance was or is carried in bulk or contained in packaged form, freight containers, portable tanks, or road and rail tanks wagons.
4. Each report shall be supplemented as necessary by any relevant information requested by a recipient of the report of deemed appropriate by the person sending the report.
5. Any of the persons referred to in article 8, paragraph 1, of this Protocol shall:
   (a) supplement as far as possible the initial report, as necessary, with information concerning further developments; and
   (b) comply as fully as possible with requests from affected States for additional information.

Protocol for the Protection of the Mediterranean Sea against Pollution from

Land-Based Sources The Contracting Parties to the Present Protocol,

Being Patties to the Convention for the Protection of the Mediterranean Sea against Pollution, adopted at Barcelona on 16 February 1976,

Desirous of implementing article 4, paragraph 2, and, articles 8 and 15 of the said Convention,

Noting the rapid increase of human activities in the Mediterranean Sea Area, particularly in the fields of industrialization and urbanization, as well as the seasonal increase in the coastal population due to tourism,

Recognizing the danger posed to the marine environment and to human health by pollution from land-based sources and the serious problems resulting therefrom in many coastal waters and river estuaries of the Mediterranean Sea, primarily due to the release of untreated, insufficiently treated or inadequately disposed domestic or industrial discharges,

Recognizing the differences in levels of development between the coastal States, and taking account of the economic and social imperatives of the developing countries, Determined to take in close co-operation the necessary measures to protect the Mediterranean Sea against pollution from land-based sources.

Have Agreed as follows:

Article 1

The Contracting Parties to this Protocol (hereinafter referred to as "the Parties") shall take all appropriate measures to prevent, abate, combat and control pollution of the mediterranean Sea Area caused by discharges from river, coastal establishments or outfalls, or emanating from any other land-based sources within their territories.

Article 2

For the purposes of this Protocol;

(a) "The Convention" means the Convention for the Protection of the Mediterranean Sea against Pollution, adopted at Barcelona on 16 February 1976;

(b) "Organization" means the body referred to in article 13 of the Convention;

(c) "Freshwater limit" means the place in watercourses where, at low tides and in a period of low freshwater flow, there is an appreciable increase in salinity due to the presence of sea water.

Article 3

The area to which this Protocol applies (hereinafter referred to as the "Protocol Area') shall be:

(a) The Mediterranean Sea Area as defined in article 1 of the Convention;

(b) Waters on the landward side of the baselines from which the breadth of the territorial sea is measured and extending, in the case of watercourses, up to the freshwater limit;

(c) Saltwater marshes communicating with the sea.

Article 4

1. This Protocol shall apply:

(a) To polluting discharges reaching the Protocol Area from land-based sources within the territories of the Parties, in particular:

— Directly, from outfalls discharging into the sea or through coastal disposal;

— Indirectly, through rivers canals or other watercourses, including underground watercourses, or through run-off;

To pollution from land-based sources transported by the atmosphere, were conditions to be defined in an additional annex to this Protocol and opted by the Parties in conformity with the provisions of article 17 of the mention.

This Protocol shall also apply to polluting discharges from fixed man off-shore structures which are under the jurisdiction of a Party and which serve purposes other than exploration and exploitation of mineral resources of the continental shelf and the sea-bed and its sub-soil.

Article 5

1. The Parties undertake to eliminate pollution of the Protocol Area from land-based sources by substances listed in annex I to this Protocol.

2. To this end they shall elaborate and implement, jointly or individually, as appropriate, the necessary programmes and measures.
3. These programmes and measures shall include, in particular, common emission standards and standards for use.
4. The standards and the time-tables for the implementation of the programmes and measures aimed at eliminating pollution from land-based sources shall be fixed by the Parties and periodically reviewed, if necessary every two years, for each of the substances listed in annex I, in accordance with the provisions of article 15 of this Protocol.

Article 6

1. The Parties shall strictly limit pollution from land-based sources in the Protocol Area by substances or sources listed in annex II to this Protocol.
2. To this end they shall elaborate and implement, jointly or individually, as appropriate, suitable programmes and measures.
3. Discharges shall be strictly subject to the issue, by the competent national authorities, of an authorization taking due account of the provisions of annex III to this Protocol.

Article 7

1. The Parties shall progressively formulate and adopt, in co-operation with the competent international organizations, common guidelines and, as appropriate, standards or criteria dealing in particular with:
   (a) The length, depth and position of pipelines for coastal outfalls, taking into account, in particular, the methods used for pretreatment of effluents;

(b) Special requirements for effluents necessitating separate treatment;

(c) The quality of sea water used for specific purposes that is necessary for the protection of human health, living resources and ecosystems;

(d) The control and progressive replacement of products, installations and industrial and other processes causing significant pollution of the marine environment;

(e) Specific requirements concerning the quantities of the substances listed in annexes I and II discharged, their concentration in effluents and methods of discharging them.

2. Without prejudice to the provisions of article 5 of this Protocol, such common guidelines, standards or criteria shall take into account local ecological, geographical and physical characteristics, the economic capacity of the Parties and their need for development, the level of existing pollution and the real absorptive capacity of the marine environment.

3. The programmes and measures referred to in articles 5 and 6 shall be adopted by taking into account, for their progressive implementation, the capacity to adapt and reconvert existing installations, the economic capacity of the Parties and their need for development.

Article 8

Within the framework of the provisions of, and the monitoring programmes provided for in, article 10 of the Convention, and if necessary in co-operation with the competent international organizations, the Parties shall carry out at the earliest possible date monitoring activities in order;

(a) Systematically to assess, as far as possible, the levels of pollution along their coasts, in particular with regard to the substances or sources listed in annexes I and II, and periodically to provide information in this respect;

(b) To evaluate the effects of measures taken under this Protocol to reduce pollution of the marine environment.

Article 9

In conformity with article 11 of the Convention, the Parties shall co-operate as far as possible in scientific and technological fields related to pollution from land-based sources, particularly research on inputs, pathways and effects of pollutants and on the development of new methods for their treatment reduction or elimination. To this end the Parties shall, in particular endeavour to:

(a) Exchange scientific and technical information;

(b) Co-ordinate their research programmes.

Article 10

1. The Parties shall, directly or with the assistance of competent regional or other international organizations or bilaterally, co-operate with a view to formulating and, as far as possible, implementing programmes to assistance to developing countries, particularly in the fields of science, education and technology, with a view to preventing pollution from land-based sources and its harmful effect in the marine environment.

2. Technical assistance would include, in particular, the training of scientific and technical personnel, as well as the acquisition, utilization and production by those countries of appropriate equipment on advantageous terms to be agreed upon among the Parties concerned.

Article 11

1. If discharges from a watercourse, which flows through the territories of two or more Parties or forms a boundary between them are likely to cause pollution of the marine environment of the Protocol Area, the Parties in question, respecting the provisions of this Protocol in so far as each of them is concerned, are called upon to co-operate with a view to ensuring its full application.
2. A Party shall not be responsible for any pollution originating on the territory of a non-contracting State. However, the said Party shall endeavour to co-operate with the said so as to make possible full application of the Protocol.

Article 12

1. Taking into account article 22, paragraph 1, of the Convention, when land-based pollution originating from the territory of one Party is likely to prejudice directly the interests of one or more of the other Parties, the parties concerned shall, at the request of one or more of them, undertake to enter into consultation with a view to seeking a satisfactory solution.
2. At the request of any Party concerned, the matter shall be placed on the agenda of the next meeting of the Parties held in accordance with article 14 of this Protocol; the meeting may make recommendations with a view to reaching a satisfactory solution.

Article 13

1. The Parties shall inform one another through the Organization of measures taken, of results achieved and, if the case arises, of difficulties encountered in the application of this Protocol. Procedures for the collection and submission of such information shall be determined at the meetings of the Parties.

2. Such information shall include, *inter alia:* (a)- Statistical data on the authorizations granted in accordance with article 6 of this Protocol;

   (b) Data resulting from monitoring as provided for in article 8 of this Protocol;

   (e) Quantities of pollutants discharged from their territories;

   (d) Measures taken in accordance with articles 5 and 6 of this Protocol.

Article 14

1. Ordinary meetings of the Parties shall take place in conjunction with ordinary meetings of the Contracting Parties to the Convention held pursuant to article 14 of the Convention. The Parties may also hold extraordinary meetings in -accordance with article 14 of the Convention.

2. The functions of the meetings of the Parties to this Protocol shall be, *inter alia:*

   (a) To keep under review the implementation of this Protocol and to consider the efficacy of the measures adopted and the advisability of any other measures, in particular in the form of annexes;

   (b) To revise and amend any annex to this Protocol, as appropriate;

   (c) To formulate and adopt programmes and measures in accordance with articles 5, 6 and 15 of this Protocol;

   (d) To adopt, in accordance with article 7 of this Protocol; common guidelines, standards or criteria, in any form decided upon by the Parties;

   (e) To make recommendations in accordance with article 12, paragraph 2, of this Protocol;

(f) To consider the information submitted by the Parties under article 13 of this Protocol;

(g) To discharge such other functions as may be appropriate for the application of this Protocol.

Article 15

1. The meeting of the Patties shall adopt, by a two-thirds majority, the programmes and measures for the abatement or the elimination of pollution from land-based sources which are provided for in articles 5 and 6 of this Protocol.

2. The Parties which are not able to accept a programme or measures shall inform the meeting of the Parties of the action they intend to take as regards the programme or measures concerned, it being understood that these Parties may, at any time, give their consent to the programme or measures that have been adopted.

Article 16

1. The provisions of the Convention relating to any Protocol shall apply with respect to this Protocol.

2. The rules of procedure and the financial rules adopted pursuant to article 18 of the Convention shall apply with respect to this Protocol, unless the Parties to this Protocol agree otherwise.

3. This Protocol shall be open for signature, at Athens from 17 May 1980 to 16 June 1980, and at Madrid from 17 June 1980 to 16 May 1981, by any State invited to the Conference of Plenipotentiaries of the Coastal States of the Mediterranean Region for the Protection of the Mediterranean Sea against Pollution from Land-Based Sources held at Athens from 12 May to 17 May 1980. It shall also be open until the same dates for signature by the European Economic Community and by any similar regional

economic grouping of which at least one member is a coastal State of the Mediterranean Sea Area and which exercises competence in fields covered by this Protocol.

4. This Protocol shall be subject to ratification, acceptance or approval. Instruments of ratification, acceptance or approval shall be deposited with the Government of Spain, which will assume the functions of Depositary.
5. As from 17 May 1981, this Protocol shall be open for accession by the States referred to in paragraph 3 above, by the European Economic Community and by any grouping referred to in that paragraph.
6. This Protocol shall enter into force on the thirtieth day following the deposit of at least six instruments of ratification, acceptance or approval of, or accession to the Protocol by the Parties referred to in paragraph 3 of this article.

In witness whereof the undersigned, being duly authorized by their respective Governments, have signed this Protocol.

Done at Athens on this seventeenth day of May one thousand nine hundred and eighty in a single copy in the Arabic, English, French and Spanish languages, the four texts being equally authoritative.

Annex I

A, The following substances, families and groups of substances are listed, not in order of priority, for the purposes of article 5 of this Protocol. They have been selected mainly on the basis of their toxicity

— persistence

— bioaccumulation.

1. Organohalogen compounds and substances which

may form such compounds in the marine environment.

2. Organophosphorus compounds and substances which may form such compounds in the marine environment.
3. Organotin compounds and substances which may form such compounds in the marine environment.
4. Mercury and mercury compounds.
5. Cadmium and cadmium compounds
6. Used lubricating oils.
7. Persistent synthetic materials which may float, sink or remain in suspension and which may interfere with any legitimate use of the sea.
8. Substances having proven carcinogenic, teratogenic or mutagenic properties in or through the marine environment.
9. Radioactive substances, including their wastes, when their discharges do not comply with the principles of radiation protection as defined by the competent international organizations, taking into account the protection of the marine environment.

B. The present annex does not apply to discharges which contain the substances listed in section A that are below the limits defined jointly by the Parties.

Annex II

A. The following substances, families and groups of substances, or sources of pollution, listed not in order of priority for the purpose of article 6 of this Protocol, have been selected mainly on the basis of criteria used for annex I, while taking into account the fact that they are generally less noxious or are more readily rendered harmless by natural processes and therefore generally affect more limited coastal areas.

1. The following elements and their compounds.
    1. zinc
    2. selenium
    3. tin
    4. vanadium
    5. copper
    6. arsenic
    7. barium
    8. cobalt
    9. nickel
    10. antimony
    11. beryllium
    12. thallium
    13. chromium
    14. molybdenum
    15. boron
    16. tellurium
    17. lead
    18. titanium
    19. uranium
    20. silver
2. Biocides and their derivatives not covered in annex I.
3. Organosilicon compounds and substances which may form such compounds in the marine environment, excluding those which are biologically harmless or are rapidly converted into biologically harmless substances.
4. Crude oils and hydrocarbons of any origin.
5. Cyanides and fluorides.

6. Non-biodegradable detergents and other surface-active substances.
7. Inorganic compounds of phosphorus and elemental phosphorus.
8. Pathogenic micro-organisms.
9. Thermal discharges.
10. Substances which have a deleterious effect on the taste and/or smell of products for human consumption derived from the aquatic environment, and compounds liable to give rise to such substances in the marine environment.
11. Substances which have, directly or indirectly, an adverse effect on the oxygen content of the marine environment, especially those which may cause eutrophication.
12. Acid or alkaline compounds of such composition and in such quantity that they may impair the quality of sea water.
13. Substances which, though of a non-toxic nature, may become harmful to the marine environment or may interfere with any legitimate use of the sea owing to the quantities in which they are discharged.

B. The control and strict limitation of the discharge of substances referred to in section A above must be implemented in accordance with annex III.

Annex III

With a view to the issue of an authorization for the discharge of wastes containing substances referred to in annex II or in section B of annex I of this Protocol; particular account will be taken, as the case may be, of the following factors:

A. *Characteristics and composition of the waste*

1. Type and size of waste source (e.g. industrial process).
2. Type of waste (origin, average composition).
3. Form of waste (solid, liquid, sludge, slurry).
4. Total amount (volume discharged, e.g. per year).
5. Discharge pattern (continuous, intermittent, seasonally variable, etc.).
6. Concentrations with respect to major constituents, substances listed in annex I, substances listed in annex 11, and other substances as appropriate.
7. Physical, chemical and biochemical properties of the waste.

B. Characteristics of waste constituents with respect to their harmfulness

1. Persistence physical, chemical, biological) in the marine environment.
2. Toxicity and other harmful effects.
3. Accumulation in biological materials or sediments
4. Biochemicals transformation producing harmful compounds.
5. Adverse effects on the oxygen content and balance.
6. Susceptibility to physical, chemical and biochemical changes and interaction in the aquatic environment with other sea water constituents which may produce harmful biological or other effects on any of the uses listed in section E below.

C. Characteristics of discharge site and receiving marine environment

1. Hydrographic, meteorological, geological and topographical characteristics of the coastal area.
2. Location and type of the discharge (outfall, canal, outlet, etc.) and its relation to other areas (such as

amenity areas, spawning, nursery, and fishing areas, shellfish grounds) and other discharges.

3. Initial dilution achieved at the point of discharge into the receiving marine environment.
4. Dispersion characteristics such as effects of currents, tides and wind on horizontal transport and vertical mixing.
5. Receiving water characteristics with respect to physical, chemical, biological and ecological conditions in the discharge area.
6. Capacity of the receiving marine environment to receive waste discharges without undesirable effects.

7 Convention on Civil Liability for Oil Pollution Damage Resulting From Exploration for and Exploitation of Seabed Mineral Resources (1977)

London, 1 May 1977

The States Parties to this Convention,

Conscious of the dangers of oil pollution posed by the exploration for, and exploitation of, certain seabed mineral resources,

Convinced of the need to ensure that adequate compensation is available to persons who suffer damage caused by such pollution,

Desiring to adopt uniform rules and procedures for determining questions of liability and provide adequate compensation in such cases,

Have agreed as follows:

Article 1

1. For the purposes of this Convention:

   a) "Oil" means crude oil and natural gas liquids, whether or not such oil or liquids are mixed with or present in other substances; and

b) "crude oil" includes crude oil treated to render it suitable for transmission, for example, by adding or removing certain fractions.

2. "Installation" means:

a) any well or other facility, whether fixed or mobile, which is used for the purpose of exploring for, producing, treating, storing, transmitting or regaining, control of the flow of crude oil from the seabed or its subsoil;

b) any well which has been used for the purpose of exploring for, producing or regaining control of the flow of crude oil from the seabed or its subsoil and which has been abandoned after the entry into force of this Convention for the Contracting State concerned;

c) any well which is used for the purpose of exploring for, producing or regaining control of the flow of gas or natural gas liquids from the seabed or its subsoil during the period that any such well is being drilled, including completion, or worked upon except for normal maintenance operations;

d) any well which is used for the purpose of exploring for any mineral resources other than crude oil, gas or natural gas liquids, where such exploration involves the deep penetration of the subsoil of the seabed; and

e) any facility which is normally used for storing crude oil from the seabed or its subsoil; which, or a substantial part of which, is located seaward of the low-water line along the coast as marked on large scale charts officially recognized by the Controlling State;

provided, however, that

(i) where a well or a number of wells is directly connected to a platform or similar facility, the well or wells together with such platform or facility shall constitute one installation; and

(ii) a ship as defined in the International Convention on Civil Liability for Oil Pollution Damage, done at Brussels on 29 November 1969, shall not be considered to be an installation.

3. "Operator" means the person, whether license or not, designated as operator for the purposes of this Convention by the Controlling State, or, in the absence of such designation, the person who is in overall control of the activities carried on at the installation.

4. "Controlling State" means the State Party which exercises sovereign rights for the purpose of exploring for and exploiting the resources of the seabed and its subsoil in the area in or above which the installator is situated. In the case of an installation extending over areas in which two or more States Parties exercise such rights, these States may agree which of them shall be the Controlling State.

5. "Person" means any individual or partnership or any public or private body, whether corporate or not, including a state or any of its constituent subdivisions.

6. "Pollution damage" means loss of damage outside the installation caused by contamination resulting from the escape or discharge of oil from the installation and includes the cost of preventive measures and further loss or damage outside the installation caused by preventive measures.

7. "Preventive measures" means any reasonable measures taken by any person in relation to a particular incident to prevent or minimize pollution damage with the exception of well control measures and measures taken to protect repair or replace an installation.
8. "Incident" means any occurrence, or series of occurrences having the same origin, which causes pollution damage.
9. "Special Drawing Right" means Special Drawing Right as defined by the International Monetary Fund and used for its own operations and transactions.

Article 2

This Convention shall apply exclusively to pollution damage:

a) resulting from an incident which occurred beyond the coastal low-water line at an installation under the jurisdiction of a Controlling State, and

b) suffered in the territory, including the internal waters and territorial sea, of a State Party or in the areas in which, in accordance with international law, it has sovereign rights over natural resources, and to preventive measures, whether taken, to prevent or minimize such pollution damage.

Article 3

1. Except as provided in paragraphs 3, 4 and 5 of this Article, the operator of the installation at the time of an incident shall be liable for any pollution damage resulting from the incident. When the incident consists of a series of occurrences, liability for pollution damage arising out of each occurrence

shall attach to the operator of the installation at the time of that occurrence.

2. Where an installation has more than one operator they shall be jointly and severally liable.

3. No liability for pollution damages shall attach to the operator if he proves that the damage resulted from an act of war, hostilities, civil war, insurrection, or natural phenomenon of an exceptional, inevitable and irresistible character.

4. No liability for pollution damage shall attach to the operator of an abandoned well if he proves that the incident which caused the damage occurred more than five years after the date on which the well was abandoned under the authority and in accordance with the requirements of the Controlling State. Where a well has been abandoned in other circumstances, the liability of the operator shall be governed by the applicable national law.

5. If the operator proves that the pollution damage resulted wholly or partly either from an act of omission done with intent to cause damage by the person who suffered the damage or from the negligence of that person, the operator may be exonerated wholly or partly from his liability to such person.

Article 4

1. No claim for compensation for pollution damage shall be made against the operator otherwise in accordance with this Convention.

2. No claim for compensation for pollution damage under this Convention or otherwise may be made against the servants or agents of the operator.

3. Nothing in this Convention shall prejudice the question whether the operator liable for damage in accordance with its provisions has a right of recourse.

Article 5

1. When oil has escaped or has been discharged form two or more installations, and pollution damage results therefrom, the operators of all the installations concerned, unless exonerated under Article 3, shall be jointly and severally liable for all such damage which is not reasonably separable.

2. When oil has escape or has been discharged from one installation as a result of an incident, and pollution damage results therefrom, and during the course of the incident there is a change of operator, all operators of the installation, unless exonerated under Article 3, shall be jointly and severally liable for all such damage which is not reasonably separable.

Article 6

1. The operator shall be entitled to limit his liability under this Convention for each installation and each incident to the amount of 30 million Special Drawing Rights until five years have elapsed from the date on which the Convention is opened for signature and to the amount of 40 million Special Drawing Rights thereafter.

2. Where operators of different installations are liable in accordance with paragraph 1 of Article 5 the liability of the operator of any one installation shall not for any one incident exceed any limit which may be applicable to him in accordance with the provisions of this Article and of Article 15.

3. When in the case of any one installation more than one operator is liable under this Convention, the aggregate liability of all of them in respect of any one incident shall not exceed the highest amount that could be awarded against any of them, but none of them shall be liable for an amount in excess of the limit applicable to him.

4. The operator shall not be entitled to limit his liability if it is proved that the pollution damage occurred as a result of an act or omission by the operator himself, done deliberately with actual knowledge that pollution damage would result.

5. For the purpose of availing himself of the benefit of limitation to which he may be entitled under paragraph 1 of this Article, the operator shall constitute a fund of the total sum representing the limit of his liability with court or other competent authority of any one of the States Parties in which action is brought under Article 11. A fund constituted by one of the operators mentioned in paragraph 2 of Article 3 shall be deemed to be constituted by all of them. The fund can be constituted either by depositing the sum or by producing a bank guarantee or other guarantee, acceptable under the legislation of the State Party where the fund is constituted, and considered to be adequately by the court or other competent liability.

6. The fund shall be distributed among the claimants in proportion to the amounts of their established claims.

7. If before the fund is distributed the operator or any of his servants or agents or any person providing him with insurance or other financial security, has, as a result of the incident in question, paid compensation for pollution damage, such person

shall, up to the amount he has paid, acquire by subrogation the rights which the person so compensated would have enjoyed under this Convention.

8. The right of subrogation provided for in paragraph 7 of this Article may also be exercised by a person other than those mentioned therein in respect of any amount of compensation for pollution damage which he may have paid but not only to the extent that such subrogation is permitted under the applicable national law.

9. Where the operator or any other person established that he may be compelled to pat at a later date in whole or in part any such amount of compensation, with regard to which such person would have enjoyed a right of subrogation under paragraph 7 or 8 of this Article, had the compensation been paid before the fund was distributed, the court or other competent authority of the State Party where the fund has been constituted may order that sufficient sum shall be provisionally set aside to enable such person at such later date to enforce his claim against the fund.

10. An operator who has taken preventive measures shall in respect of those measures have the same rights against the fund as any other claimant.

11. The amount referred to in paragraph 1 of this Article shall be converted into the national currency of the State Party in which the fund is constituted on the basis of the value of that currency by reference to the average, during the thirty days immediately preceding the date on which the fund is constituted, of the Special Drawing Rights as published by the International Monetary fund.

12. The insurer or other person providing financial security shall be entitled, alone or together with the operator, to constitute a fund in accordance with this Article on the same conditions and having the same effect as if it were constituted by the operator. Such a fund may be constituted even where the pollution damage occurred as a result of an act or omission by the operator himself, done deliberately with actual knowledge that pollution damage would result, but the constitution of the fund shall in that case not prejudice the rights of any claimant against the operator.

Article 7

1. Where the operator, after an incident, has constituted a fund in accordance with Article 6 and is entitled to limit his liability.

   a) no person having a claim for pollution damage arising out of that incident shall be entitled to exercise any right against any other assets of the operator in respect of such claim:

   b) the court or other competent authority of any State Party shall order the release of any property belonging to the operator which has been arrested in respect of acclaim for pollution damage arising out of that incident, and shall similarly release any bail or other security furnished to avoid such arrest.

2. Paragraph 1 of this Article shall, however, only apply if the claimant has access to the court administering the fund and the fund is actually available in respect of his claim.

Article 8

1. To cover his liability under this Convention, the operator shall be required to have and maintain

insurance or other financial security to such amount, of such type an d on such terms as the Controlling State shall specify, provided that the, amount shall not be less than 22 million Special Drawing Rights until five years have elapsed from the date on which this Convention is opened for signature and not less than 35 million Special Drawing Rights thereafter. However the Controlling State may exempt the operator wholly, or in part from the requirement to have and maintain such insurance or other financial security to cover his liability for pollution damage wholly caused by an act of sabotage or terrorism.

2. An insurance or other financial security shall not satisfy the requirements of this Article if it clan cease, for reasons other than the expiry of the period of validity of the insurance or security, before two months have elapsed from the date on which notice of its termination is given to the competent public authority of the Controlling State. The foregoing provision shall similarly apply to any modification which results in the insurance or security no longer satisfying the requirements of this Article.

3. Any claim for compensation for pollution damage may be brought direct against the insurer or other person providing financial security for the operator's liability for pollution damage. In such case the liability of the defendant shall be limited to the amount specified in accordance with paragraph 1 of this Article irrespective of the fact that the pollution damage occurred as a result of an act or omission by the operator himself, done deliberately with actual knowledge that pollution damage would result. The defendant may further avail himself of the defences other than the bankruptcy or winding up of the operator, which the operator himself would have been entitled to invoke. Furthermore, the defendant

may avail himself of the defence that the pollution damage resulted from the wilful misconduct of the operator himself, but the defendant may not avail himself of any other defence which he might have been entitled to invoke in proceedings brought by the operator against him. The defendant shall in any event have the right to require the operator to be joined in the proceedings.

4. Any sums provided by insurance or by other financial security maintained in accordance with paragraph 1 of this Article shall be available in the first place for the satisfaction of claims under this Convention.

5. Where the operator is a State Party, the operator shall not be required to maintain insurance or other financial security to cover its liability.

Article 9

1. A Committee composed of a representative of each State Party is hereby established.

2. If a State Party considers that any of the amounts currently applicable under Article 6 or 8 is no longer adequate, or is otherwise unrealistic, it may convene a meeting of the Committee to consider the matter. States which have signed this Convention but are not yet Parties will be invited to participate in the work of the Committee may recommend to the States Parties an amendment to any of the amounts if representatives of at least three quarters of the States Parties to this Convention vote in favour of such a recommendation. In making such a recommendation, the Committee shall take into account;

   a) any information concerning events causing or likely to cause pollution damage having a bearing on the objects of this Convention;

b) any information on increases and decreases occurring after the entry into force of this Convention in the costs of goods and services of the kinds involved in the treatment and remedying of marine oil spillages;

c) the availability of reliable insurance cover against the risk of liability for pollution damage.

3. Any amount recommended in accordance with paragraph 2 of this Article shall be notified by the depositary Government to all States Parties. It shall replace the amount currently applicable thirty days after its acceptance by all States Parties. A State Party which has, within six months of such notification or such other period as has been specified in the recommendation, notified the depositary Government that it is unable to accept the recommended amount, shall be deemed to have accepted, it.

4. If the recommended amount has not been accepted by all States Parties within six months, or such other period as has been notified by the depositary Government it shall, thirty days thereafter, replace the amount currently applicable as between those States Parties which have accepted it. Any other State Party may subsequently accept the recommended amount which shall become applicable to it thirty days thereafter.

5. A state acceding to this Convention shall be bound by any recommendation of the Committee which has been unanimously accepted by States Parties. Where a recommendation has not been so accepted, an acceding State shall be deemed to have accepted it unless, at the time of its accession, that State notifies the depositary Government that it does not accept such recommendation.

Article 10

Rights of compensation under this Convention shall be extinguished unless, within twelve months of the date on which the person suffering the damage knew or ought reasonably to have known of the damage, the claimant has in writing notified the operator of his claim or has brought an action in respect of it. However in no case shall an action be brought after four years from the date of the incident which caused the damage. Where the incident consists of a series of occurrences, the four year's period shall run from the date of the last occurrence.

Article 11

1. Actions for compensation under this Convention may be brought only in the courts of any State Party where pollution damage was suffered as a result of the incident or in the courts of the, Controlling State. For the purpose of determining where the damage was suffered, damage suffered in an area in which, in accordance with international law, a State has sovereign rights over natural resources shall be deemed to have been suffered in that State.

2. Each State Party shall ensure that its courts possess the necessary jurisdiction to entertain such actions for compensation.

3. After the fund has been constituted in accordance with Article 6, the courts of the State Party in which the fund is constituted shall be exclusively competent to determine all matters relating to the apportionment and distribution of the fund.

Article 12

1. Any judgement given by a court with jurisdiction in accordance with Article 11, which is enforceable in the State of origin where it is no longer subject to ordinary d in any State Party, forms of review, shall be recognize except:

a) where the judgement was obtained by fraud; or

b) where the defendant was not given reasonable notice and a fair opportunity to present his case.

2. A judgement recognized under paragraph 1 of this Article shall be enforceable in each State Party as soon as formalities required in that State have been complied with. The formalities shall not permit the merits of the case to be re-opened, nor a reconsideration of the applicable law.

Article 13

Where a State Party is the operator, such State shall be subject to suit in the jurisdictions set forth in Article 11 and shall waive all defences based on its status as a sovereign State.

Article 14

No liability shall arise under this Convention for damage caused by a nuclear incident:

a) if the operator of a nuclear installation is liable-for such damage under either the Paris Convention of 29 July 1960 on Third Party Liability in the Field of Nuclear Energy or the Vienna Convention of 21 May 1963 on Civil Liability for Nuclear Damage, or if the operator of a nuclear ship is liable for such damage under the Brussels Convention of 25 May 1962 on the Liability of Operators of Nuclear Ships; or

b) if the operator of a nuclear installation or the operator of a nuclearship is liable for such damage by virtue of a national law governing the liability for such damage, provided that such law is in all respects as favourable to persons who may suffer damage as, in the case of the operator of nuclear installation, either the Paris or the Vienna Convention or, in the case of the operator of a nuclear ship, the Brussels Convention.

Article 15

1. This Convention shall not prevent a State from providing for unlimited liability or a higher limit of liability than that currently applicable under Article 6 for pollution damage caused by installations for which it is the Controlling State and suffered in that State or in another State Party; provided however that in so doing it shall not discriminate on the basis of nationality. Such provision may be based on the principle of reciprocity.

2. The courts of each State Party shall apply the law of the Controlling State in order to determine whether the operator is entitled under the provisions of this Article and paragraph 1 of Article 6 to limit his liability and, if so, the amount of such liability.

3. Nothing in this Article shall affect the amount of compensation available for pollution damage suffered in States Parties in respect of which the provision made in accordance with paragraph 1 of this Article does not apply.

4. For the purposes of this Article, pollution damage suffered in a State Party means pollution damage suffered in the territory of that State or in the areas in which, in accordance with international law, it has sovereign rights over natural resources.

Article 16

This Convention shall be open for signature at London from 1 May 1977 until 30 April 1978 by the States invited to participate in the Intergovernmental Conference on the

Convention on Civil Liability for Oil Pollution Damage from Offshore Operations, held there from 20 October to 31 October 1975 and from 13 December to 17 December 1976, and shall thereafter be open for accession by such States.

Article 17

The Convention shall be subject to ratification acceptance or approval.

Article 18

The States Parties may unanimously invite to accede to this Convention other States which have coastline on the North Sea, the Baltic Sea or that part of the Atlantic Ocean to the north of 36° North latitude.

Article 19

The instruments of ratification, acceptance, approval and accession shall be deposited with the Government of the United Kingdom of Great Britain and Northern Ireland.

Article 20

1. This Convention shall enter into force on the ninetieth day following the date of deposit of the fourth instrument of ratification, acceptance, approval or accession.
2. For each State ratifying, accepting, approving or acceding to the Convention after the deposit of the fourth instrument, the Convention shall enter into force on the ninetieth day after deposit by such State of its instrument.

Article 21

A State Party may denounce this Convention at any time by means of a notice in writing addressed to the depositary Government. Any such denunciation shall take effect twelve months after the date on which the depositary Government has received such notice, or at such later date as may be specified in the notice.

Article 22

1. Any State may, at the time of ratification, acceptance, approval or accession or at any later date, declare by means of a notice in writing addressed to the depositary Government that this Convention shall apply to all or any of the territories for whose international relations it is responsible, provided that they are situated within the area defined in Article 18.

2. Such declaration shall take effect on the ninetieth day after its receipt by the depositary Government or, if on such date the Convention has not yet entered into force, from the date of its entry into force.

3. Each State Party which has made a declaration in accordance with paragraph 1 of this Article 21, denounce this Convention in relation to all or any of the territories concerned.

Article 23

Any State Party may, after having obtained the agreement of at least one-third of the States Parties convene a Conference of States Parties for the revision or amendment of this Convention.

Article 24

No reservation may be made to this Convention.

Article 25

The depositary Government shall inform the States referred to in Article 16 and the acceding States:

a) of signatures to this Convention, of the deposit of instruments of ratification, acceptance, approval or accession, of the receipt of notices in accordance with Article 22, and of the receipt of notices of denunciation;

b) of the date on which the Convention will enter in to force; and

c) of the recommendations of the Committee convened under Article 9, of the acceptances and non-acceptances of such recommendations, and of the dates on which these recommendations take effect.

Article 26

The original of the Convention, of which the English and French texts are equally authentic, shall be deposited with the Government of the United Kingdom of Great Britain and Northern Ireland, which shall send certified copies thereof to the States referred to in Article 16 and the acceding States and which upon its entry into force, shall transmit a certified copy to the Secretariat of the United Nations for registration and publication in accordance with Article 102 of the Charter of the United Nations.

In witness whereof the undersigned. being duly authorised thereto, have signed this Convention.

Done at London this First Day of May, 1977.

8 UNCLOS III Settlement of Disputes and Advisory Opinions (1982)

Article 186

Sea-Bed Disputes Chamber of the International Tribunal for the Law of the Sea

The establishment of the Sea-Bed Disputes Chamber and the manner in which it shall exercise its jurisdiction shall be governed by the provision of this section, of Part XV and of Annex VI

The Sea-Bed Disputes Chamber shall have jurisdiction under this Part and the Annexes relating thereto in disputes with respect to activities in the Area falling within the following categories:

(a) disputes between State Parties concerning the interpretation or application of this Part and the Annexes relating thereto;

(b) disputes between a State Party and the Authority concerning:

(i) acts or omissions of the Authority or of a State Party alleged to be in violation of this Part of the Annexes relating thereto or of rules, regulations and procedures of the Authority adopted in accordance therewith; or

(ii) acts of the Authority alleged to be in excess of jurisdiction or a misuse of power;

(c) disputes between parties to a contract, being States Parties, the Authority or the Enterprise, state enterprises and natural or juridical persons referred to in article 153, paragraph 2(b), concerning:

(i) the interpretation or application of a relevant contract or a plan of work; or

(ii) acts or omissions of a party to the contract relating to activities in the Area and directed to the other party or directly affecting its legitimate interests;

(d) disputes between the Authority and a prospective contractor who has been sponsored by a State as provided in article 153, paragraph 2 (b), and has duly fulfilled the conditions referred to in Annex III, article 4, paragraph 6, and article 13, paragraph 2, concerning the refusal of a contract or a legal issue arising in the negotiation of the contract;

(e) disputes between the Authority and a State Party, a state enterprise or a natural or juridical person sponsored by a State Party as provided for in article 153, paragraph 2(b), where it is alleged that the Authority has incurred liability as provided in Annex III, article 22;

(f) any other disputes for which the jurisdiction of the Chamber is specifically provided in this Convention.

Article 188

Submission of disputes to a special chamber of the International Tribunal for the Law of the Sea or an ad hoc chamber of the Sea-Bed Disputes Chamber or to binding commercial arbitration

1. Disputes between States Parties referred to in article 187, sub-paragraph (a), may be submitted:

   (a) at the request of the parties to the dispute, to a special chamber of the International Tribunal for the Law of the Sea to be formed in accordance with Annex VI, articles 15 and 17; or

   (b) at the request of any party to the dispute, to an ad hoc chamber of the Sea-Bed Disputes Chamber to be formed in accordance with Annex VI, article 36.

2. (a) Disputes concerning the interpretation or application of a contract referred to in article 187, sub-paragraph (c) (i), shall be submitted, at the request of any party to the dispute, to binding commercial arbitration, unless the parties otherwise agree. A commercial arbitral tribunal to which the dispute is submitted shall have no jurisdiction to decide any question of interpretation of this Convention. When the dispute also involves a question of the interpretation of Part XI and the Annexes relating thereto, with respect to activities in the Area, that question shall be referred to the Sea-Bed Disputes Chamber for a ruling.

   (b) If, at the commencement of or in the course of such arbitration, the arbitral tribunal determines, either at the request of any party to the dispute or *proprio motu*, that its decision depends upon a ruling of the Sea-Bed Disputes Chamber, the arbitral tribunal shall refer such question to the Sea-Bed Disputes Chamber for such ruling. The arbitral tribunal shall then proceed to render its award in conformity with the ruling of the Sea-Bed Disputes Chamber.

   (c) In the absence of a provision in the contract on the arbitration procedure to be applied in the dispute, the arbitration shall be conducted in accordance wit h the UNCITRAL Arbitration Rules or such other arbitration rules as may be prescribed in the rules, regulations and procedures of the Authority, unless the parties to the dispute otherwise agree.

Article 189

Limitation on jurisdiction with regard to decisions of the Authority

The Sea-Bed Disputes Chamber shall have no jurisdiction with regard to the exercise by the Authority of its discretionary powers in accordance with this Part; in no case shall it substitute its discretion for that of the Authority. Without prejudice to article 191, in exercising its jurisdiction pursuant to article 187, the Sea-Bed Disputes Chamber shall not pronounce itself on the question of whether any rules, regulations and procedures of the Authority are in conformity with this Convention, nor declare invalid any such rules, regulations and procedures. Its jurisdiction in this regard shall be confined to deciding claims that the application of any rules, regulations and procedures of the Authority in individual cases would be in conflict with the contractual obligations of the parties to the dispute or their obligations under this Convention, claims concerning excess of jurisdiction or misuse of power, and to claims for damages to be paid or other remedy to be given to the party concerned for the failure of the other party to comply with its contractual obligations or its obligations under this Convention.

Article 190

Participation and appearance of sponsoring State Parties in proceedings

1. If a natural or juridical person is a party to a dispute referred to in article 187, the sponsoring State shall

be given notice thereof and shall have the right to participate in the proceedings by submitting written or oral statements.

2. If an action is brought against a State Party by .a natural or juridical person sponsored by another State Party in a dispute referred to in article 187, sub-paragraph (c), the respondent State may request the State sponsoring that person to appear in the proceedings on behalf of that person. Failing such appearance, the respondent State may arrange to be represented by a juridical person of its nationality.

Article 191

Advisory opinions

The Sea-Bed Disputes Chamber shall give advisory opinions at the request of the Assembly or the Council on legal questions arising within the scope of their activities. Such opinions shall be given as a matter of urgency.

Part XV

Settlement of Disputes

General Provisions

Article 279

Obligation to settle disputes by peaceful means

States Parties shall settle any dispute between them concerning the interpretation or application of this Convention by peaceful means in accordance with Article 2, paragraph 3, of the Charter of the United Nations and, to this end, shall seek a solution by the means indicated in Article 33, paragraph 1, of the Charter.

Article 280

Settlement of disputes by any peaceful means chosen by the parties

Nothing in this Part impairs the right of any States Parties to agree at any time to settle a dispute between them concerning the interpretation or application of this Convention by any peaceful means of their own choice.

Article 281

Procedure where no settlement has been reached by the parties

1. If the States Parties which are parties to a dispute concerning the interpretation or application of this Convention have agreed to seek settlement of the dispute by a peaceful means of their own choice, the procedures provided for in this Part apply only where no settlement has been reached by recourse to such means and the agreement between the parties does not exclude any further procedure.
2. If the parties have also agreed on a time-limit, paragraph 1 applies only upon the expiration of that time-limit.

Article 282

Obligations under general, regional or bilateral agreements

If the State Parties which are parties to a dispute concerning the interpretation or application of this Convention have agreed, through a general, regional or bilateral agreement or otherwise, that such dispute shall, at the request of any party to the dispute, be submitted to a procedure that entails a binding decision, that procedure shall apply in lieu of the procedures provided for in this Part, unless the parties to the dispute otherwise agree.

Article 283

Obligation to exchange views

1. When a dispute arises between States Parties concerning the interpretation or application of this Convention, the parties to the dispute shall proceed expeditiously to an exchange of views regarding its settlement by negotiation or other peaceful means.

2. The parties shall also proceed expeditiously to an, exchange of views where a procedure for the settlement of such a dispute has been terminated without a settlement or where a settlement has been reached and the circumstances require consultation regarding the manner of implementing the settlement

Article 284

Conciliation

1. A State Party which is a party to a dispute concerning the interpretation or application of this Convention may invite the other party or parties to submit the dispute to conciliation in accordance with the procedure under Annex V, section 1, or another conciliation procedure.

2. If the invitation is accepted and if the parties agree upon the conciliation procedure to be applied, any party may submit the dispute to that procedure.

3. If the invitation is not accepted or the parties do not agree upon the procedure, the conciliation proceedings shall be deemed to be terminated.

4. Unless the parties otherwise agree, when a dispute has been submitted to conciliation, the proceedings may be terminated only in accordance with the agreed conciliation, procedure.

Article 285

Application of this section to disputes submitted pursuant to Part XI

This section applies to any dispute which pursuant to Part XI, section 5, is to be settled in accordance with procedures provided for in this Part. If an entity other than a State Party is a party to such a dispute, this section applies mutatis mutandis.

Compulsory Procedures Entailing Binding Decisions

Article 286

Application of procedures, under this section

Subject to section 3, any dispute concerning the interpretation or application of this Convention shall, where no settlement has been reached by recourse to section 1, be submitted at the request of any party to the dispute to the court or tribunal having jurisdiction under this section.

Article 287

Choice of procedure

1. When signing, ratifying or acceding to this Convention or at any time thereafter, a State shall be free to choose, by means of a written declaration, one or more of the following means for the settlement of disputes concerning the interpretation or application of this Convention:
   (a) the International Tribunal for the Law of the Sea established in accordance with Annex VI;
   (b) the International Court of Justice;
   (c) an arbitral tribunal constituted in accordance with Annex VII;
   (d) a special arbitral tribunal constituted in accordance with Annex VIII for one or more of the categories of disputes specified therein.
2. A declaration made under paragraph 1 shall not affect or be affected by the obligation of a State Party to accept the jurisdiction of the Sea-Bed Disputes Chamber of the International Tribunal for the Law of the Sea to the extent and in the manner provided for in Part XI, section 5.

3. A State Party, which is a party to a dispute not covered by a declaration in force, shall be deemed to have accepted arbitration in accordance with Annex VII.

4. If the parties to a dispute have accepted the same procedure for the settlement of the dispute, it may be submitted only to that procedure, unless the parties otherwise agree.

5. If the parties to a dispute have not accepted the same procedure for the settlement of the dispute, it may be submitted only to arbitration in accordance with Annex VII, unless the parties otherwise agree.

6. A declaration made under paragraph 1 shall remain in force until three months after notice of revocation has been deposited with the Secretary-General of the United Nations.

7. A new declaration, a notice of revocation or the expiry of a declaration does not in any way affect proceedings pending before a court or tribunal having jurisdiction under this article, unless the parties otherwise agree.

8. Declarations and notices referred to in this article shall be deposited with the Secretary-General of the United Nations who shall transmit copies thereof to the States Parties.

Article 288

Jurisdiction

1. A court or tribunal referred to in article 287 shall have jurisdiction over any dispute concerning the interpretation or application of this Convention which is submitted to it in accordance with this Part.

2. A court or tribunal referred to in article 287 shall also have jurisdiction over any dispute concerning

the interpretation or application of an international agreement related to the purposes of this Convention, which is submitted to it in accordance with the agreement.

3. The Sea-Bed Disputes Chamber of the International Tribunal for the Law of the Sea established in accordance with Annex VI, and any other chamber or arbitral tribunal referred to in Part XI section 5, shall have jurisdiction in any matter which is submitted to it in accordance therewith.

4. In the event of a dispute as to whether a court or tribunal has jurisdiction, the matter shall be settled by decision of that court or tribunal.

Article 289

Experts

In any dispute involving scientific or technical matter, a court or tribunal exercising jurisdiction under this section may, at the request of a party or *proprio motu*, select in consultation with the parties no fewer than two scientific or technical experts chosen preferably from the relevant list prepared in accordance with Annex VIII, article 2, to sit with the court or tribunal but without the right to vote.

Article 290

Provisional measures

1. If a dispute has been duly submitted to a court or tribunal which considers that prima facie it has jurisdiction under this Part or Part XI, section 5, the court or tribunal may prescribe any provisional measures which it considers appropriate under the circumstances to preserve the respective rights of the parties to the dispute or to prevent serious harm to the marine environment, pending the final decision.

2. Provisional measures may be modified or revoked as soon as the circumstances justifying them have changed or ceased to exit.

3. Provisional measures may be prescribed, modified or revoked under this article only at the request of a party to the dispute and after the parties have been given and opportunity to be heard.

4. The court or tribunal shall forthwith give notice to the parties to the dispute, and to such other States Parties as it considers appropriate, of the prescription, modification or revocation of provisional measures.

5. Pending the constitution of an arbitral tribunal to which a dispute is being submitted under this section, any court or tribunal agreed upon by the parties or, failing such agreement within two weeks from the date of the request for provisional measures, the International Tribunal for the Law of the Sea or, with respect to activities in the Area, the Sea-Bed Disputes Chamber, may prescribe, modify or revoke provisional measures in accordance with this article if it considers that prima facie the tribunal which is to be constituted would have jurisdiction and that the urgency of the situation so requires.

6. The parties to the dispute shall comply promptly with any provisional measures prescribed under this article.

Article 291

Access

1. All the dispute settlement procedures specified in this Part shall be open to States Parties.

2. The dispute settlement procedures specified in this

Part shall be open to entities other than States Parties only as specifically provided for in this Convention.

Article 292

Prompt release of vessels and crews

1. Where the authorities of a State Party have detained a vessel flying the flag of another State Party, and it is alleged that the detaining State has not complied with the provisions of this Convention for the prompt release of the vessel or its crew upon the posting of a reasonable bond or their financial security, the question of release from detention may be submitted to any court or tribunal agreed upon by the parties or, failing such agreement within 10 days from the time of detention to a court or tribunal accepted by the detaining State under article 287 or to the International Tribunal for the Law of the Sea, unless the parties otherwise agree.

2. The application may be made only by or on behalf of the flag State of the vessel.

3. The court or tribunal shall deal without delay with the application for release and shall deal only with the question or release, without prejudice to the merits of any case before the appropriate domestic forum against the vessels, its owner or its crew. The authorities of the detaining State remain competent to release the vessels or its crew at any time.

4. Upon the posting of the bond or other financial security determined by the court or tribunal, the authorities of the detaining State shall comply promptly with the decision of the court or tribunal concerning the release of the vessel or its crew.

Article 293

Applicable law

1. A court or tribunal having jurisdiction under this section shall apply this Convention and other rules of international law not incompatible with this Convention.

2. Paragraph 1 does not prejudice the power of the court or tribunal having jurisdiction under this section to decide a case *ex aequo et bono,* if the parties so agree.

Article 294

Preliminary proceedings

1. A court or tribunal provided for in article 287 to which an application is made in respect of a dispute referred to in article 297 shall determine at the request of a party or may determine proprio motu, whether the claim constitutes an abuse of legal process or whether prima facie it is well founded. If the court or tribunal determines that the claim constitutes an abuse of legal process or is prima facie unfounded, it shall take no further action in the case.

2. Upon receipt of the application, the court or tribunal shall immediately notify the other party or parties of the application, and shall fix a reasonable time-limit within which they may request it to a determination in accordance with paragraph 1.

3. Nothing in this article affects the right of any party to a dispute to make preliminary objections in accordance with the applicable rules of procedure.

Article 295

Exhaustion of local remedies

Any dispute between States Parties concerning the interpretation or application of this Convention may be submitted to the procedures provided for in this section only after local remedies have been exhausted where this is required by international law.

Article 296

Finally and binding force of decisions

1. Any decision rendered by a court or tribunal having jurisdiction under this section shall be final and shall be complied with by all the parties to the dispute.
2. Any such decision shall have not binding force except between the parties and in respect of that particular dispute.

Limitations and Exceptions to Applicability of Section 2

Article 297

Limitations on applicability of section 2

1. Disputes concerning the interpretation or application of this Convention with regard to the exercise by a coastal State of its sovereign rights or jurisdiction provided for in this Convention shall be subject to the procedures provided for in section 2 in the following cases:
   (a) when it is alleged that a coastal State has acted in contravention of the provisions of this Convention in regard to the freedoms and rights of navigation, overflight or the laying of submarine cables and pipelines or in regard to other internationally lawful uses of the sea specified in article 58;
   (b) when it is alleged that a State in exercising the aforementioned freedoms, rights or uses has acted in contravention of this Convention or of laws or regulations adopted by the coastal State in conformity with this Convention and other

rules of international law not incompatible with this Convention; or

(c) when it is alleged that coastal State has acted in contravention of specified international rules and standards for the protection and preservation of the marine environment which are applicable to the coastal State and which have been established by this Convention or through a competent international organization or diplomatic conference in accordance with this Convention.

2. (a) Disputes concerning the interpretation or application of the provisions of this Convention with regard to marine scientific research shall be settled in accordance with section 2, except that the coastal State shall not be obliged to accept the submission to such settlement of any dispute arising out of:

(i) the exercise by the coastal State of a right or discretion in accordance with article 246; or

(ii) a decision by the coastal State to order suspension or cessation of a research project in accordance with article 253.

(b) A dispute arising from an allegation by the researching State that with respect to a specific project the coastal State is not exercising its rights under articles 246 and 253 in a manner compatible with this Convention shall be submitted, at the request of either party, to conciliation under Annex V, section 2, provided that the conciliation commission shall not call in question the exercise by the coastal State of its discretion to designate specific areas as referred to in article 246, paragraph 6, or of its discretion to withhold consent in accordance with article 246, paragraph 5.

3. (a) Disputes concerning the interpretation or application of the provisions to this Convention with regard to fisheries shall be settled in accordance with section 2, except that the coastal State shall not be obliged to accept the submission to such settlement of any dispute relating to its sovereign rights with respect to the living resources in the exclusive economic zone or their exercise, including its discretionary powers for determining the allowable catch, its harvesting capacity, the allocation of surpluses to other States and the terms and conditions established in its conservation and management laws and regulations.

(b) Where no settlement has been reached by recourse to section 1 of this Part, a dispute shall be submitted to conciliation under Annex V, section 2, at the request of any party to the dispute, when it is alleged that:

(i) a coastal State has manifestly failed to comply with its obligations to ensure through proper conservation and management measures that the maintenance of the living resources in the exclusive economic zone is not seriously endangered;

(ii) a coastal State has arbitrarily refused to determine, at the request of another State, the allowable catch and its capacity to harvest living resources with respect to stocks which that other State is interested in fishing; or

(iii) a coastal State has arbitrarily refused to allocate to any State, under articles 62, 69 and 70 under the terms and conditions established by the coastal State consistent with this Convention, the whole or part of the surplus it has declared to exist.

(c) In no case shall the conciliation commission substitute its discretion for that of the coastal State.

(d) The report of the conciliation commission shall be communicated to the appropriate international organizations.

(e) In negotiating agreements pursuant to articles 69 and 70, States Parties, unless they otherwise agree, shall include a clause on measures which they shall take in order to minimize the possibility of a disagreement concerning the interpretation or application of the agreement, and on how they should proceed if a disagreement nevertheless arises.

Article 298

Optional exceptions to applicability of section 2

When signing, ratifying or acceding to this Convention or at any time thereafter, a State may, without prejudice to the obligations arising under section 1, declare in writing that it does not accept any one or more of the procedures provided for in section 2 with respect to one or more of the following categories of disputes:

(a) (i) disputes concerning the interpretation or application of articles 15, 74 and 83 relating to sea boundary delimitations, or those involving historic bays or titles, provided that a State having made such a declaration shall, when such a dispute arises subsequent to the entry into force of this Convention and where no agreement within a reasonable period of time is reached in negotiations between the parties, at the request of any party to the dispute, accept submission of the matter to conciliation under Annex V, section 2, and provided further that any dispute that necessarily involves the concurrent consideration of any unsettled dispute concerning sovereignty or other rights over continental or insular land territory shall be excluded from such submission;

(ii) after the conciliation commission has presented its report, which shall state the reasons on which it is based, the parties shall negotiate an agreement on the basis of that report; if these negotiations do not result in an agreement, the parties shall, by mutual consent, submit the question to one of the procedures provided for in section 2, unless the parties otherwise agree;

(iii) this sub-paragraph does not apply to any sea boundary dispute finally settled by an agreement between the parties, or to any such dispute which is to be settled in accordance with a bilateral or multilateral agreement binding upon those parties;

(b) disputes concerning military activities, including military activities by government vessels and aircraft engaged in non-commercial service, and disputes concerning law enforcement activities in regard to the exercise of sovereign rights or jurisdiction excluded from the jurisdiction of a court or tribunal under article 297, Paragraph 2 or 3;

(c) dispute in respect of which the Security Council of the United Nations is exercising the functions assigned to it by the Charter of the United Nations, unless the Security Council decides to remove the matter from its agenda or calls upon the parties to settle it by the means provided for in this Convention.

2. A State Party which has made a declaration under paragraph 1 may at any time withdraw it, or agree to submit a dispute excluded by such declaration to any procedure specified in this Convention.

3. A State Party which has made a declaration under paragraph 1 shall not be entitled to submit any dispute falling within the excepted category of disputes to any procedure in this Convention as against another State Party, without the consent of that Party.

4. If one of the States Parties has made a declaration under paragraph 1 (a), any other State Party may submit any dispute falling within an excepted categories against the declarant party to the procedure specified in such declaration.

5. A new declaration, or the withdrawal of a declaration, does not in any way affect proceedings pending before a court or tribunal in accordance with this article, unless the parties otherwise agree.

6. Declarations and notices of withdrawal of declarations under this article shall be deposited with the Secretary-General of the United Nations, who shall transmit copies thereof to the States Parties.

Article 299

Right of the parties to agree upon a procedure

1. A dispute excluded under article 297 or excepted by a declaration made under article 298 from the dispute settlement procedures provided for in section 2 may be submitted to such procedures only by agreement of the parties to the dispute.

2. Nothing in this section impairs the right of the parties to the dispute to agree to some other

procedure for the settlement of such dispute or to reach an amicable settlement.

Annex VI

Statute of the International Tribunal for the Law of the Sea

Article 1

General provisions

1. The International Tribunal for the Law of the Sea is constituted and shall function in accordance with the provisions of this Convention and this Statute.

2. The seat of the Tribunal shall be in the Free and Hanseatic City of Hamburg in the Federal Republic of Germany.

3. The Tribunal may sit and exercise its functions elsewhere wherever it considers this desirable.

4. A reference of a dispute to the Tribunal shall be governed by the provisions of Parts XI and XV.

Organization of the Tribunal

Article 2

Composition

1. The Tribunal shall be composed of a body of 21 independent members, elected from among persons enjoying the highest reputation for fairness and integrity and of recognized competence in the field of the law of the sea.

2. In the Tribunal as a whole the representation of the principal legal systems of the world and equitable geographical distribution shall be assured.

Article 3

Membership

1. No two members of the Tribunal may be nationals

of the same State. A person who for the purposes of membership in the Tribunal could be regarded as a national of more than one State shall be deemed to be a national of the one in which the ordinarily exercises civil and political rights.

2. There shall be no fewer than three members from each geographical group as established by the General Assembly of the United Nations.

Article 4

Nominations and elections

1. Each State Party may nominate not more than two persons having the qualifications prescribed in article 2 of this Annex. The members of the Tribunal shall be elected from the list of persons thus nominated.

2. At least three months before the date of the election, the Secretary-General of the United Nations in the case of the first election and the Registrar of the Tribunal in the case of subsequent elections shall address a written invitation to the States Parties to submit their nominations for members of the Tribunal within two months. He shall prepare a list in alphabetical order of all the persons thus nominated, with an indication of the State Parties which have nominated them, and shall submit it to the States Parties before the seventh day of the last month before the date of each election.

3. The first election shall be held within six months of the date of entry into force of this Convention.

4. The members of the Tribunal shall be elected by secret ballot. Elections shall be held at a meeting of the States Parties convened by the Secretary-General of the United Nations in the case of the first election

and by a procedure agreed to by the States Parties in the case of subsequent elections. Two thirds of the States Parties shall constitute a quorum at that meeting. The persons elected to the Tribunal shall be those nominees who obtain the largest number of votes and two-thirds majority of the States Parties present and voting, provided that such majority includes a majority of the States Parties.

Article 5

Term of office

1. The members of the Tribunal shall be elected for nine years and may be re-elected; provided, however, that of the members elected at the first election, the terms of seven members shall expire at the end of three years and the terms of seven more members shall expire at the end of six years.

2. The members of the Tribunal whose terms are to expire at the end of the above-mentioned initial periods of three and six years shall be chosen by lot to be drawn by the Secretary-General of the United Nations immediately after the first election.

3. The members of the Tribunal shall continue to discharge their duties until their places have been filled. Though replaced, they shall finish any proceedings which they may have begun before the date of their replacement.

4. In the case of the resignation of a member of the Tribunal, the letter of resignation shall be addressed to the President of the Tribunal. The place becomes vacant on the receipt of that letter.

Article 6

Vacancies

1. Vacancies shall be filled by the same method as that

laid down for the first election, subject to the following provisions: the Registrar shall, within one month of the occurrence of the vacancy, proceed to issue the invitations provided for the article 4 of this Annex, and the date of the election shall be fixed by the President of the Tribunal after consultation with the States Parties.

2. A member of the Tribunal elected to replace a member whose term of office has not expired shall hold office for the remainder of his predecessor's term.

Article 7

Incompatible activities

1. No member of the Tribunal, may exercise any political or administrative function, or associate actively with or be financially interested in any of the operations of any enterprises concerned with the exploration for or exploitation of the resources of the sea or the sea-bed or other commercial use of the sea or the sea-bed.

2. No member of the Tribunal may act as agent, counsel or advocate in any case.

3. Any doubt on these points shall be resolved by decision of the majority of the other members of the Tribunal present.

Article 8

Conditions relating to participation of members in a particular case

1. No member of the Tribunal may participate in the decision of any case in which he has previously taken part as agent, counsel or advocate for one of the parties, or as a member of a national or international court or tribunal, or in any other capacity.

2. If, for some special reason, a member of the Tribunal considers that he should not take part in the decision of a particular case, he shall so inform the President of the Tribunal.
3. If the President considers that for some special reason one of the members of the Tribunal should not sit in a particular case, he shall give him notice accordingly.
4. Any doubt on these points shall be resolved by decision of the majority of the other members of the Tribunal present.

Article 9

Consequence of ceasing to fulfil required conditions

If, in the unanimous opinion of the other members of the Tribunal, a member has ceased to fulfil the required conditions, the President of the Tribunal shall declare the seat vacant.

Article 10

Privileges and immunities

The members of the Tribunal, when engaged on the business of the Tribunal shall enjoy diplomatic privileges and immunities.

Article 11

Solemn declaration by members

Every member of the Tribunal shall, before taking up his duties, make a solemn declaration in open session that he will exercise his powers impartially and conscientiously.

Article 12

President, Vice-President and Registrar

1. The Tribunal shall elect its President and Vice-President for three years; they may be re-elected.

2. The Tribunal shall appoint its Registrar and may provide for the appointment of such other officers as may be necessary.

3. The President and the Registrar shall reside at the seat of the Tribunal.

Article 13

Quorum

1. All available members of the Tribunal shall sit; a quorum of 11 elected members shall be required to constitute the Tribunal.

2. Subject to article 17 of this Annex, the Tribunal shall determine which members are available to constitute the Tribunal for the consideration of a particular dispute, having regard to the effective functioning of the chambers as provided for in articles 14 and 15 of this Annex.

3. All disputes and applications submitted to the Tribunal shall be heard and determined by the Tribunal, unless article 14 of this Annex applies, or the parties request that it shall be dealt with in accordance with article 15 of this Annex.

Article 14

Sea-Bed Disputes Chamber

A Sea-Bed Disputes Chamber shall be established in accordance with the provisions of section 4 of this Annex. Its jurisdiction, powers and functions shall be as provided for in Part XI section 5.

Article 15

Special chambers

1. The Tribunal may form such chambers, composed of three or more of its elected members, as it considers necessary for dealing with particular categories of disputes.

2. The Tribunal shall form a chamber for dealing with a particular dispute submitted to it if the parties so request. The composition of such a chamber shall be determined by the Tribunal with the approval of the parties.

3. With a view to the speedy dispatch of business, the Tribunal shall form annually a chamber composed of five of its elected members which may hear and determine disputes by summary procedure. Two alternative members shall be selected for the purpose of replacing members who are unable to participate in a particular proceeding.

4. Disputes shall be heard and determined by the chambers provided for in this article if the parties so request.

5. A judgment given by any of the chambers provided for in this article and in article 14 of this Annex shall be considered as rendered by the Tribunal.

Article 16

Rules of the Tribunal

The Tribunal shall frame rules for carrying out its functions. In particular it shall lay down rules of procedure.

Article 17

Nationality of members

1. Members of the Tribunal of the nationality of any of the parties to a dispute shall retain their right to participate as members of the Tribunal.

2. If the Tribunal, when hearing a dispute, includes upon the bench a member of the nationality of one of the parties, any other party may choose a person to participate as a member of the Tribunal.

3. If the Tribunal, when hearing a dispute, does not include upon the bench a member of the nationality of the parties, each of those parties may choose a person to participate as a member of the Tribunal.

4. This article applies to the chambers referred to in articles 14 and 15 of this Annex. In such cases, the President, in consultation with the parties, shall request specified members of the Tribunal forming the chamber, as many as necessary, to give place to the members of the Tribunal of the nationality of the parties concerned, and failing, suck or if they are unable to be present, to the members specially chosen by the parties.

5. Should there be several parties in the same interest, they shall, for the purpose of the preceding provisions, be considered as one party only. Any doubt on this point shall be settled by the decision of the Tribunal.

6. Members chosen in accordance with paragraphs 2, 3 and 4 shall fulfil the conditions required by articles 2, 8 and 11 of this Annex. They shall participate in the decision on terms of complete equality with their colleagues.

Article 18

Remuneration of members

1. Each elected member of the Tribunal shall receive an annual allowance and, for each day on which he exercises his functions, a special allowance, provided that in any year the total sum payable to any member as special allowance shall not exceed the amount of the annual allowance.

2. The President shall receive a special annual allowance.

3. The Vice-President shall receive a special allowance for each day on which he acts as President.

4. The members chosen under article 17 of this Annex, other than elected members of the Tribunal, shall receive compensation for each day on which they exercise their functions.

5. The salaries, allowances and compensation shall be determined from time to time at meetings of the States Parties, taking into account the work load of the Tribunal. They may not be decreased during the term of office.

6. The salary of the Registrar shall be determined at meetings of the States Parties on the proposal of the Tribunal.

7. Regulations adopted at meetings of the States Parties shall determine the conditions under which retirement pensions may be given to members of the Tribunal and to the Registrar, and the conditions under which members of the Tribunal and Registrar shall have their travelling expenses refunded.

8. The salaries, allowances, and compensation shall be free of all taxation.

Article 19

Expenses of the Tribunal

1. The expenses of the Tribunal shall be borne by the States Parties and by the Authority on such terms and in such a manner as shall be decided at meetings of the States Parties.

2. When an entity other than a State Party or the Authority is a party to a case submitted to it, the Tribunal shall fix the amount which that party is to contribute towards the expenses of the Tribunal.

Competence

Article 20

Access to the Tribunal

1. The Tribunal shall be open to States Parties.
2. The Tribunal shall be open to entities other than States Parties in any case expressly provided for in Part XI or in any case submitted pursuant to any other agreement conferring jurisdiction on the Tribunal which is accepted by all the parties to that case.

Article 21

Jurisdiction

The jurisdiction of the Tribunal comprises all disputes and all applications submitted to it in accordance with this Convention and all matters specifically provided for in any other agreement which confers jurisdiction on the Tribunal.

Article 22

Reference of disputes subject to other agreements

If all the parties to a treaty or convention already in force and concerning the subject-matter covered by this Convention so agree, any disputes concerning the interpretation or application of such treaty or convention may, in accordance with such agreement, be submitted to the Tribunal.

Article 23

Applicable law

The Tribunal shall decide all disputes and applications in accordance with article 293.

Procedure

Article 24

Institution of Proceedings

1. Disputes are submitted to the Tribunal, as the case may be, either by notification of a special agreement or by written application, addressed to the Registrar. In either case, the subject of the dispute and the parties shall be indicated.
2. The Registrar shall forthwith notify the special agreement or the application to all concerned.
3. The Registrar shall also notify all States Parties.

Article 25

Provisional measures

1. In accordance with article 290, the Tribunal and its Sea-Bed Disputes Chamber shall have the power to prescribe provisional measures.
2. If the Tribunal is not in session or a sufficient number of members is not available to constitute a quorum, the provisional measures shall be prescribed by the chamber of summary procedure formed under article 15, paragraph 3, of this Annex. Notwithstanding article 15, paragraph 4, of this Annex, such provisional measures may be adopted at the request of any party to the dispute. They shall be subject to review and revision by the Tribunal.

Article 26

Hearing

1. The hearing shall be under the control of the President or, if he is unable to preside, of the Vice President. If neither is able to preside, the senior judge present of the Tribunal shall preside.

2. The hearing shall be public, unless the Tribunal decides otherwise or unless the parties demand that the public be not admitted.

Article 27

Conduct of case

The Tribunal shall make orders for the conduct of the case, decide the form and time in which each party must conclude its arguments, and make all arrangements connected with the taking of evidence.

Article 28

Default

When one of the parties does not appear before the Tribunal or fails to defend its case, the other party may request the Tribunal to continue the proceedings and make its decision. Absence of a party or failure of a party to defend its case shall not constitute a bar to the proceedings. Before making its decision, the Tribunal must satisfy itself not only that it has jurisdiction over the dispute, but also that the claim is well founded in fact and law.

Article 29

Majority for decision

1. All questions shall be decided by a majority of the members of the Tribunal who are present.

2. In the event of an equality of votes, the President or the member of the Tribunal who acts in his place shall have a casting vote.

Article 30

Judgment

1. Judgment shall state the reasons on which it is based.

2. It shall contain the names of the members of the Tribunal who have taken part in the decision.

3. If the judgment does not represent in whole or in part the unanimous opinion of the members of the Tribunal, and member shall be entitled to deliver a separate opinion.

4. The judgment shall be signed by the President and by the Registrar. It shall be read in open court, due notice having been given to the parties to the dispute.

Article 31

Request to intervene

1. Should a State Party consider that it has an interest of a legal nature which may be affected by the decision in any dispute, it may submit a request to the Tribunal to be permitted to intervene.
2. It shall be for the Tribunal to decide upon this request.
3. If a request to intervene is granted, the decision of the Tribunal in respect of the dispute shall be binding upon the intervening State Party in so far as it relates to matters in respect of which that State Party intervened.

Article 32

Right to intervene in cases of interpretation or application

1. Whenever the interpretation or application of this Convention is in question, the Registrar shall notify all States Parties forthwith.

2. Whenever pursuant to articles 21 or 22 of this Annex the interpretation or application of an international agreement is in question, the Registrar shall notify all the parties to the agreement.

3. Every party referred to in paragraphs 1 and 2 has the right to intervene in the proceedings; if it uses this right, the interpretation given by the judgment will be equally binding upon it.

Article 33

Finally and binding force of decisions

1. The decision of the Tribunal is final and shall be complied with by all the parties to the dispute.

2. The decision shall have no binding force except between the parties in respect of that particular dispute.

3. In the event of dispute as to the meaning or scope of the decision, the Tribunal shall construe it upon the request of any party.

Article 34

Costs

Unless otherwise decided by the Tribunal, each party shall bear its own costs.

Sea-Bed Disputes Chamber

Article 35

Composition

1. The Sea-Bed Disputes Chamber referred to in article 14 of this Annex shall be composed of 11 members, selected by a majority of the elected members of the Tribunal from among them.

2. In the selection of the members of the Chamber, the representation of the principal legal systems of the world and equitable geographical distribution shall be assured. The Assembly of the Authority may adopt recommendations of a general nature relating to such representation and distribution.

3. The members of the Chamber shall be selected every three years and may be selected for a second term.

4. The Chamber shall elect its President from among its members, who shall serve for the term for which the Chamber has been selected.

5. If any proceedings are still pending at the end of any three-year period for which the Chamber has been selected, the Chamber shall complete the proceedings in its original composition.

6. If a vacancy occurs in the Chamber, the Tribunal shall select a successor from among its elected members, who shall hold office for the remainder of his predecessor's term.

7. A quorum of seven of the members selected by the Tribunal shall be required to constitute the Chamber.

Article 36

Ad hoc chambers

1. The Sea-Bed Disputes Chamber shall form an ad hoc chamber, composed of three of its members, for dealing with a particular dispute submitted to it in accordance with article 188, paragraph 1 (b). The composition of such a chamber shall be determined by the Sea-Bed Disputes Chamber with the approval of the parties.

2. If the parties do not agree on the composition of an ad hoc chamber, each party to the dispute shall appoint one member, and the third member shall be appointed by them in agreement. If they disagree, or if any party fails to make an appointment, the President of the Sea-Bed Disputes Chamber shall promptly make the appointment or appointments from among its members, after consultation with the parties.

3. Members of the ad hoc chamber must not be in the service of, or nationals or, any of the parties to the dispute.

Article 37

Access

The Chamber shall be open to the States Parties, the Authority and the other entities referred to in Part XI, section 5.

Article 38

Applicable law

In addition to the provisions of article 293, the Chamber shall apply:

(a) the rules, regulations and procedures of the Authority adopted in accordance with the Convention; and

(b) the terms of contracts concerning activities in the Area in matters relating to those contracts.

Article 39

Enforcement of decisions of the Chamber

The decisions of the Chamber shall be enforceable in the territories of the States Parties in the same manner as judgment or orders or the highest court of the State Party in whose territory the enforcement is sought.

Article 40

Applicability of other sections of this Annex

1. The other sections of this Annex which are not incompatible with this section apply to the Chamber.

2. In the exercise of its functions relating to advisory opinions, the Chamber shall be guided by the provisions of this Annex relating to procedure

before the Tribunal to the extent to which it recognizes them to be applicable.

Amendments

Article 41

Amendments

1. Amendments to this Annex, other than amendments to section 4, may be adopted only in accordance with article 313 or by consensus at a conference convened in accordance with this Convention.

2. Amendments to section 4 may be adopted only in accordance with article 314.

3. The Tribunal may purpose such amendments to this Statue as it may consider necessary, by written communications to the States Parties for their consideration in conformity with paragraphs 1 and 2.

9 Montreal Guidelines on the Protection of Marine Environment (1985)

Introduction

This set of guidelines is addressed to Governments with a view to assisting them in the process of developing appropriate bilateral, regional and multilateral agreements and national legislation for the protection of the marine environment against pollution from land-based sources. They have been prepared on the basis of common elements and principles drawn from relevant existing agreements and drawing upon experience already gained through their preparation and implementation, Principal among these agreements are the United Nations Convention on the Law of the Sea (Part XII), the Paris Convention for the Prevention of Marine Pollution from Land-Based Sources, the Helsinki Convention on the Protection of Marine Environment of the Baltic Sea Area, and the Athens Protocol for the Protection of the Mediterranean Sea against Pollution from Land-Based Sources.

These guidelines are suggested as a broad framework for the development of similar agreements in these regions ere such agreements are called for; for the guidance Governments in areas which may not presently be covered by any regional agreements; and for the preparation in the longer term, should the need arise, of global convention on pollution from land-based sources signed to strengthen international institutional arrangements to ensure the harmonization and application of global and regional rules, criteria, standards and recommended practices and procedures and to review the effectiveness of measures taken.

The guidelines are of a recommendatory nature. They are presented as a checklist of basic provisions rather than a model agreement, from which Governments may select, adapt of elaborate, as appropriate, to meet the needs of specific regions. They are without prejudice to the elaboration of cross-sectoral guidelines/principles within the framework of the programme for the development and periodic review of environmental law, as recommended by the UNEP Ad Hoc Meeting of Senior Government Official Expert in Environmental Law.

Definitions

For the purposes of these guidelines:

(a) "Pollution" means the introduction by man, directly or indirectly, of substances or energy into the marine environment which results or is likely to result in such deleterious effects as harm to living resources and marine ecosystems, hazards to human health, hindrance to marine activities, including fishing and other legitimate uses of the sea, impairment of quality for use of sea water and reduction of amenities.

(b) "Land-based sources" means:

(i) Municipal, industrial or agricultural sources, both fixed and mobile, on land, discharges from which reach the marine environment, in particular: From the coast, including from outfalls discharging directly into the marine environment and through run-off;

Through rivers, canals or other water-courses, including underground water-courses; and via the atmosphere.

(ii) Sources of marine pollution from activities conducted on offshore fixed or mobile facilities within the limits of national jurisdiction, save to the extent that these sources are governed by appropriate international agreements.

(c) "Marine environment" means the maritime area extending, in the case of water-courses up to fresh water limit and including inter-tidal zones and salt-water marshes;

(d) "Freshwater limit" means the place in watercourses where, at low tide and in a period of low freshwater flow, there is an appreciable increase in salinity due to the presence of sea-water.

Basic obligation

States have the obligation to protect and preserve the marine environment. In exercising their sovereign right to exploit their natural resources, all States have the duty to prevent, reduce and control pollution of the marine environment.

Discharges affecting other States or areas beyond the limits of national jurisdiction

States have the duty to ensure that discharges from land-based sources within their territories do not cause pollution to the marine environment of other States or of areas beyond the limits of national jurisdiction.

Adoption of measures against pollution from land-based sources

States should adopt, individually or jointly, and in accordance with their capabilities, all measures necessary to prevent, reduce and control pollution from land-based sources, including those designed to minimize to the fullest possible extent the release of toxic, harmful or noxious substances, especially those which are persistent, into the marine environment. States should ensure that such measures take into account internationally agreed rules, criteria, standards and recommended practices and procedures.

In taking measures to prevent, reduce and control pollution from land-based sources, States should refrain, in accordance with international law, from unjustifiable interference with activities carried out by other States in the exercise of their sovereign rights and in pursuance of their duties in conformity with internationally agreed rules, criteria, standards and recommended practices and procedures.

Co-operation on a global, regional or bilateral basis

States should undertake, as appropriate, to establish internationally agreed rules, criteria, standards and recommended practices and procedures to prevent, reduce and control pollution from land-based sources, with a view to co-ordinating their policies in this connection, particularly at the local and regional level. Such rules, criteria, standards and recommended practices and procedures should take into account local ecological, geographical and physical characteristics, the economic capacity of States and their need for sustainable development and environmental protection, and the assimilative capacity of the marine environment, and should be reviewed from time to time as necessary.

States not bordering on the marine environment should co-operate in preventing, reducing and controlling pollution of the marine environment originating or partially originating from releases within their territory into or reaching water basins or water-courses flowing into the marine environment or via the atmosphere. To this and, States concerned should as far as possible and, as appropriate, in co-operation with competent international organizations, take necessary measures to prevent, reduce and control pollution of the marine environment from land-based sources.

If discharges of a water-course which flows through the territories of two or more States or forms a boundary between them are likely to cause pollution of the marine environment, the States concerned should cooperate in taking necessary measures to prevent, reduce and control such pollution.

Duty not to transfer or transform pollution from land-based sources

In taking measures to prevent, reduce and control pollution from land-based sources, States have the duty to act so as not to transfer directly or indirectly, damage or hazards from one area to another or transform such pollution into another type of pollution.

Specially protected areas

States should consistent with international law, take all appropriate measures, such as the establishment of marine sanctuaries and reserves, to protect certain areas to the fullest possible extent from pollution, including that from land-based sources, taking into account the relevant provisions of Annex 1.

States should, as practicable, undertake to develop, jointly or individually, environmental quality objectives for specially protected areas, conforming with the intended uses, and strive to maintain or ameliorate existing conditions by comprehensive environmental managements practices.

Scientific and technical co-operation

States should co-operate, directly and/or through

competent international organizations, in the fields of science and technology related to pollution from land-based sources, and exchange data and other scientific information for the purpose of preventing, reducing and controlling such pollution, taking into account national regulations regarding the protection of confidential information. They should, in particular, undertake to develop and co-ordinate to the fullest possible extent their national research programmes and to co-operate in the establishment and implementation of regional and other international research programmes.

Assistance to developing countries

States should, directly and/or through competent international organizations, promote programmes of assistance to developing countries in the fields education, environmental and pollution awareness training, scientific research , transfer of technology and know-how, for the purpose of improving the capacity the developing countries to prevent, reduce and control pollution from land-based sources and to assess its effect on the marine environment.

Such assistance should include:

(a) Training of Scientific and technical personnel;
(b) Facilitation of the participation of developing countries in relevant international programmes
(c) Acquisition, utilization, maintenance an production by those countries of appropriate equipment; and
(d) Advice on, and development of, facilities for education, training, research, monitoring and other programmes.

States should directly and/or through competent international organizations, promote programmes of assistance to developing countries for the establishment, as necessary, of infrastructure for the effective implementation of applicable internationally agreed rules, criteria, standards and recommended practices and procedures related to the protection of the marine environment against pollution from land-based sources, including the provision of expert advice on the development of the necessary legal and administrative measures.

Development of a comprehensive environmental management approach

States should undertake to develop, as far as practicable, a comprehensive environmental management approach to the prevention, reduction and control of pollution from land-based sources, taking into account relevant existing programmes at the bilateral, regional or global level and the provisions of Annex 1. Such a comprehensive approach should include the identification of desired and attainable water-use objectives for the specific marine environments.

Monitoring and data management

States should endeavour to establish directly or, whenever necessary through competent international organizations, complementary or joint programmes for monitoring, storage and exchange of data, based, when possible, on compatible procedures and methods, taking into account relevant existing programmes at the bilateral, regional or global level and the provisions of Annex III, in order to:

(a) Collect data on natural conditions in the region concerned as regards its physical, biological and chemical characteristics;

(b) Collect data on input of substances or energy that causes or potentially causes pollution emanating from land-based sources, including information on the distribution of sources and the quantities introduced to the region concerned;

(c) Assess systematically the levels of pollution along their coasts emanating from land-based sources and the fate and effects of pollution in the region concerned; and

(d) Evaluate the effectiveness of measures in meeting the environmental objectives for specific marine environments.

Environmental assessment

States should assess the potential effects/impacts, including possible transboundary effects/impacts, of proposed major projects under their jurisdiction or control, particularly in coastal areas, which may cause pollution from land-based sources, so that appropriate measures may be taken to prevent or mitigate such pollution.

Development of control strategies

States should develop, adopt and implement Programmes and measures for the prevention, reduction and control of pollution from land-based sources. They should employ an appropriate control strategy or combination of control strategies, taking into account relevant international or national experience.

States should, as appropriate, progressively formulate, and adopt, in co-operation with competent international organizations, standards based on marine quality or on emissions, as well as recommended practices and procedures, taking into account the provisions.

Where appropriate, states should undertake to establish priorities for action, based on lists of substances from which pollution should be eliminated and of substances from which pollution should be strictly limited on the basis of their toxicity, persistence, bioaccumulation and other criteria.

Pollution emergencies arising from land-based sources

States and, as appropriate, competent international organizations, should take all necessary measures for preventing and dealing with marine pollution emergencies from land-based sources, however caused, and for reducing or eliminating damage or the threat of damage therefrom. To this end states should, as appropriate, individually or jointly, develop and promote a national and international contingency plans for responding to incidents of pollution from land-based sources and should co-operate with one another and, whenever necessary, through competent international organizations.

Notification, information exchange and consultation

Whenever releases originating or likely to originate from land-based sources within the territory of a State are likely to cause pollution to the marine environment of one or more other States or of areas beyond the limits of national jurisdiction, that State should immediately notify such other State or States, as well as competent international organizations, and provide them with timely information that will enable them, where necessary, to take appropriate action to prevent, reduce and control such pollution. Furthermore, consultations deemed appropriate by States concerned should be undertaken with a view to preventing, reducing and controlling such pollution.

National law and procedures

Each State should adopt and implement national laws and regulations for the protection and preservation of the marine environment against pollution from land-based sources, taking into account internationally agreed rules, criteria, standards and recommended practices and procedures, and take appropriate measures to ensure .compliance with such laws and regulations.

Paragraph 1 is without prejudice to the right of States to take more stringent measures nationally or in co-operation with each other to prevent, reduce and control pollution from land-based sources under their jurisdiction control.

Each State should, on a reciprocal basis, grant equal access to and non-discriminatory treatment in its courts, tribunals and administrative proceedings to persons in other States who are or may be affected by pollution from land-based sources under its jurisdiction control.

Liability and compensation for pollution damage emanating from land-based sources

States should ensure that recourse is available in accordance with their legal systems for prompt and adequate compensation or other relief in respect of damage caused by pollution of the marine environment by natural or juridical persons under their jurisdiction.

To this end, States should formulate and adopt appropriate procedures for the determination of liability for damage resulting from pollution from land-based sources.. Such procedures should include measures for addressing damage caused by releases of a significant scale or by the substances referred to in guideline 13, paragraph 3.

Implementation reports

States should report, as appropriate, to other States concerned, directly or through competent international organizations, on measures taken, on results achieved and, if the case arises, on difficulties encountered in the implementation of applicable internationally agreed rules, criteria, standards and recommended practices and procedures, to this end, States designates national authorities as focal points for the reporting of such measures, results and difficulties.

Institutional arrangements

States should ensure that adequate institutional arrangements are made at the appropriate regional or global level, for the purpose of achieving the objectives of these guidelines, and in particular for promoting the formulation, adoption and application of international rules, criteria, standards and recommended practices and procedures, and for monitoring the condition of the marine environment.

The function of such institutional arrangements should include:

(a) Periodic assessment of the state of the specific marine environment concerned;

(b) Formulation and adoption, as appropriate, of a comprehensive environmental management approach consistent with the provisions of guideline 7 and 10;,

(c) Adoption, review and revision, as necessary of the lists referred to in guideline 13;

(d) Development and adoption, as appropriate, of programmes and measures -consistent with the provisions of guidelines 10 and 13;

(e) Consideration, where necessary, of the reports and information submitted in accordance with guidelines 15 and 18;

(f) Recommendation of appropriate measures to be taken for the prevention, reduction and control of pollution from land-based sources, such as assistance to developing countries, the strengthening of regional mechanisms of cooperation, consideration of aspects of transboundary pollution, and the difficulties encountered in the implementation of agreed rules; and

(g) Review of the implementation of relevant

internationally agreed rules, criteria,' standards and recommended practices and procedures, and of the efficacy of the measures adopted and advisability of any other measures.

Annex I

Strategies for protecting, preserving and enhancing strategic the quality of the marine environment

Introduction

In controlling marine pollution from land-based ,sources, and overall approach to the uses and the natural values of the marine environment should be taken, while 11 considering the needs of populations and industries waste disposal. It is important to note that for many of Waste, the use of the marine environment is only e option among several. However, in some instances, marine disposal may be a feasible alternative. This ,document describes a number of strategies which can be employed to protect the marine environment against Pollution from land-based sources and, where- necessary, restore areas that have been affected. The goal is to protect the marine ecosystem by maintaining its quality within acceptable levels as determined on the basis of scientific, institutional, social and economic factors, It should be recognized that there are many activities competing to derive benefits from the marine environment, None of these activities, save the perpetuation of a marine ecosystem as a vital component of global life support,. should be regarded as having guaranteed rights. Compromise and consideration of all alternatives must always be considered. Consequently, in the course of the decision-making process determining the use of a particular sector of the marine environment, social, economic and political factors, as well as natural environmental factors must be taken into account.

Once decision makers have determined the desired

present, interim and long terms uses, and associated objectives for a water body, a number of control strategies may be employed to achieve those objectives. Flexibility will be an important consideration in the strategies or regulatory instruments implemented for various water bodies, reflecting their different environmental capacities and other properties and differences in regional socioeconomic conditions. The principal strategies in use are based on marine quality standards, on emission standards and on environmental planning. Experience shows that a combination of strategies is often needed. Practical constraints may prevent full implementation of a strategy based on quality standards. Where such an approach cannot be fully implemented, other strategies should be employed.

Control strategies

Pollution control strategies in use have been categorized according to:

— those based on marine environmental quality standards,

— those based on emission standards,

— those based on environmental planning.

Priorities for control are often established by the classification of substances into a black and a grey list substances are assessed according to the criteria described in Annex II. States undertake to eliminate pollution 41 those substances in the black list and strictly to limit pollution by those in the grey list.

Strategies based on marine quality standards

Such strategies relate directly to quality of water, biota or sediments that must be maintained for a desired level of quality and intended use. Several applications of such quality-based strategies exist.

Direct derivation from quality objectives

Technical assessments are conducted to determine the maximum allowable inputs that will ensure the desired levels of environmental quality are met. The assessments consider the fates and affects of various contaminants, amounts of input, and the existing natural characteristics of the relevant marine ecosystem. Numerical standards are then established to which concentrations measured in the receiving environment may be compared. They are usually more restrictive than numbers derived from the technical assessment to allow for monitoring and enforcement capabilities and safety requirements. They may apply to water, sediment, fish or the tissues, health or community composition of organisms in the marine ecosystem.

Monitoring is required to detect changes and compliance with the standards. Changes in the items monitored, after adjustment for natural fluctuation, may signal a need further to reduce inputs and vary existing standards and controls.

No change above ambient

Standards are set based on existing levels which Must not be exceeded. This strategy is employed in situations where the aim is to prevent any increase in prevailing specific contaminant levels. It is an interim strategy to allow time to develop a solid scientific base on which more precise quality criteria may be employed for a specific use. It does not imply an existing state of the environment that is satisfactory, nor does it eliminate the need for its improvement.

Dilution

Some contaminants discharged at the source are assumed to attenuate as they spread, from that source. Dynamic characteristics of the receiving environment are employed to determine rate and level of dilution. Standards are derived from measured parameters taken at given distances from the discharging source. The strategy may accept short-term or local excess of a potential pollution at the source of discharge. Application is generally used with effluent that is considered biodegradable, and avoided where scientific evidence suggests that the effluent may accumulate in a given receiving environment.

Loading allocations

These impose priority of control on the larger sources in consideration of the most cost-effective solution. Allowable discharges are measured in terms of the total allowable for an entire receiving environment regardless of specific site quality. Application is suited to relatively self-contained receiving environments such as lagoons and semi-enclosed bodies of water. It allows flexibility of contaminant output in that certain sources may emit more than adjacent ones as long as loading limits are not exceeded. All these strategies may employ criteria for water, air or sediment quality, as well as criteria related to specific marine life. Receiving environment quality standards are most prevalent for uses, e.g. swimming, direct harvesting of fish for human consumption, where sound scientific criteria exist to determine levels of harm. Emissions of potential pollutants are usually controlled to ensure that the desired quality is achieved. If the quality needs to be upgraded, additional controls are placed on allowable emissions.

Strategies based on emission standards

These strategies may be based on:

— a general principle to control pollution,
— achievable technology,
— distribution of control costs,

— enforceability.

They differ from strategies based on marine quality in that the standards set are not primarily determined by the level of contaminant in the environment.

Technology-based standards.

These standards are usually applied on a sectoral basis, thus providing a means of imposing similar costs .Across a particular sector. Alternatively, they may be ,determined on a case-by-case basis. The standards will need to be reviewed periodically the light of developing technology.

Standards may be based on:

Best practicable technology

This reflects the application of demonstrable and sound treatment technology or spectrum of technologies which affordable by the sector concerned.

Best available technology

This reflects state-of-art technology in use in contaminant control. In general, the standards set would reflect a more stringent level of control as compared to best practicable technology. Application is generally for control of emissions of the most noxious substances to protect a sensitive environmental use.

As low as reasonably achievable

This is mainly applied to radionuclides and is based on the principle of "optimization". This, as defined by the international Commission on Radiological Protection, requires radiation doses to be kept to levels that are "reasonable achievable", by technological improvements and by suitable choice among alternative options. "Reasonably achievable" takes into account both the ease with which the technology can be applied and the balance between the benefits, in terms of dose reduction, and social and economic costs of its application.

In a situation where stringent protection of sensitive marine environment is deemed appropriate, consideration may be given to the denial of any release of a contaminant to the environment.

Uniform regional emission standards

Such standards are usually applied in situations where there are existing pollution problems of a similar nature and there is urgent need to reduce pollution. They do not give primary consideration to the nature of sources, their economic base, or the receiving environment.

Planning strategies

Planning strategies allow an approach to the management and protection of particular environments which may involve restrictions on, or modification of, activities and sites as well as discharges.

Activity management

Certain activities are deemed inappropriate or inconsistent with the value or use of an environment. Consideration should be given to whether the activity is essential, and if so, whether it can be accommodated elsewhere or in a different manner.

Use designation

Use of the receiving environment is the determining factor for pollution control standards as well, as the basis for regulations or guidelines affecting other activities. For example, if the desire is to maintain or develop a shellfish harvest (a socio-economic decision) then quality standards and uses are developed with, this in mind.

The application may result from the perceived threat to an established economic base, or cultural value, or, a conscious effort to change the existing use of the receiving environment.

Environmental assessments of activities

Siting of any activity significantly affecting the marine environment is subject to a comprehensive analysis and assessment of :

— The ecological characteristics of the receiving environment,
— the direct and indirect potential effects/impacts of the activity on the environment; and, as appropriate,
— the direct and indirect potential effects/impacts on the environment of any reasonable alternative to the activity.

Regional planning

Plans are drawn up for particular region, taking into account socio-economic and ecological factors, which are then used as a basis for development.

Coastal zone management

The strategy employs planning capabilities to make best use of the coastal zone.

It is not uses or source specific but area specific assessed as components of a coastal zone planning is based on regional socio-economic and ecological considerations. Zoning and other land-use restrictions or modifications are major regulatory tools. Many state employ the use of regional planning authorities of councils, given the task to manage overall resource planning within A particular coastal area.

Watershed or drainage basin planning

This strategy acknowledges that a large proportion of pollution enters the marine environment via watercourses. It does not necessarily account for inputs via the atmosphere, though air management areas have also been employed for control purposes.

Through consideration of socio-economic an environmental factors utilizing a drainage system as the boundary limit, the desired uses and level of quality that can be attained for any given marine water body are determined.

Pollution via watercourses is controlled through regulation of points and diffuse sources of such pollution within the given watershed.

Specially protected areas

This strategy involves the identification of unique or pristine areas, rare or fragile ecosystems, critical habitats and the habitat of dipleted, threatened or endangered species and other forms of marine life.

Those areas to be protected or preserved from pollution, including that from land-based sources, are selected on the basis of a comprehensive evaluation of factors, including conservational ecological, recreational, aesthetic and *scientific values.*

State should notify an appropriate international organization of the establishment of and any modification to such areas, with a view to that data being included in an inventory of specially protected areas.

Control Instruments

This section outlines the various types of mechanism which can be invoked to implement control strategies.

Regulations

Regulations are developed pursuant to establishing legislation and can exist in form such as:

Emissions standards (air/water)

Standards based on best practicable technology, best available technology, geographical area, etc.,

Environmental quality standards

Standards for the receiving environment which vary according to its intended use.

Guidelines of practice

These are descriptions of practices and abatement technologies that may be developed to meet the pollution control needs of various point and non-point sources. They provide a listing of basic requirements that may be implemented or adopted by industry or local authorities.

Permits

Legislation may require a discharger to have a permit to satisfy the requirements for the release of pollutants. These requirements can be based on standards in the form of emission control regulations, guidelines, codes of practice or specific requirement derived from environmental quality standard prescribed to protect the receiving environment.

Equipment standards certification

Environmental considerations may be incorporated directly in association with particular equipment. To this end, the equipment, or configuration of equipment may be designed, manufactured, tested and certified to comply with the requirements for source releases of pollulants.

Product controls

If a particular substance, or assemblage of substances in the form of a commercial product is deemed to be of environmental significance, a restriction on the product in the form of production, use, as well as export/import may be implemented.

Planning restrictions

Under planning law or practice, restrictions 'may be placed on the use of certain land.

Economic measures

These may take a variety of forms, e.g. tax incentive, subsidies and effluent charges. To be effective, the incentive offered must be strong enough or the charge levied high enough to persuade the discharger of user that it is in his own financial interest to limit his discharge or use of the substance concerned.

Factors Influencing Choice of Strategies and Control Instruments

There is a wide range of strategies and control instruments which can be utilized by a state either individually or in combination to address pollution, of the marine environment from land-based sources. A number of factors may influence such a choice. In general terms, they may be categorized as economic, scientific/technical or social/cultural/political.

Economic

— General economic conditions and trends (deficit, balance of trade, inflation etc.),

— Availability of public financing,

— Availability of external funding,

— Unemployment

— Economic viability of various sectors,

— Polluter-pays principle,

— Availability of institutions and infrastructure.

Scientific/Technical

Availability-accessibility of scientific data, including.

— Physical characteristic affecting flushing and mixing,

— Natural nutrient cycles and geochemical, cycles,
— Biological processes and nature of communities.

Availability/ accessibility of technology, including.

— Basic information on industry types, total effluent releases, and specific data on waste stream constituents.
— Availability of expertise,
— capability for monitoring,
— existing engineering infrastructure
— Experience with implementation of strategies or instruments elsewhere,
— Sensitivity of ecosystems to be affected,
— Climatic consideration,
— Current level of pollution of the receiving environment and identified trends, in municipal agricultural and industrial waste releases.

Social/cultural/political

— Infrastructure,
— Existing and proposed uses of the marine environment,
— Political realities,
— Social/cultural awareness of the population,
— Perception of environmental, social and cultural values.

Annex II

Classification of substances

Introduction

Substances may be classified into a black list of those substances pollution from which should be eliminated and grey list of those substances pollution from which should be strictly limited and reduced.

The basic criteria to be taken into account in allocating substances to one of these lists are:

(a) persistence,

(b) toxicity to bio-accumulation,

(c) tendency to bio-accumulation.

These criteria are not necessarily of equal importance for a particular substance or group of substances. Other factors such as location and quantities of the discharge may need to be considered.

Black List

Substances may be included in this list:

(a) because they are not readily degradable or rendered harmless by natural processes; and

(b) because they may either:

 (i) give rise to dangerous accumulation of harmful material in the food chain, or

 (ii) endanger the welfare of living organism causing undesirable changes in the marine ecosystems, or

 (iii) interfere seriously with the harvesting of set foods or with other legitimate uses of the sea and

(c) because it is considered that pollution by these substances necessitates urgent action.

The substances that fulfil criteria may include:

— certain organic biocides (e.g. organohalogen compounds and substances which may form such compounds in the marine environment);

— Persistent hydrocarbons of petroleum origin;

— Certain metals and their compounds (e.g. mercury);

— Persistent synthetic materials which may seriously interfere with any legitimate use of the sea;

— Radioactive materials;
— Substances in respect of which it has been proved that they possess carcinogenic properties in or via the acquatic environment;
— Materials in whatever form (e.g. solids, liquids, semi-liquids, gases, or in a living state) produced for biological and chemical warfare.

Grey List

Substances may be included in this list because, although exhibiting similar characteristics to the substances in the black list and requiring strict control, they seem less noxious or are more readily rendered harmless by natural processes. The substances to which this may apply include:

— Organic biocides not included in the black list;
— Hydrocarbons of petroleum origin and their derivatives not included in the black list;
— Certain elements and their compounds (e.g. fluorides and cyanides);
— Organic and synthetic organic materials, other those included in the black list, which are likely to produce harmful effects on marine organisms or to make ~edible marine organisms unpalatable, as well as chemicals which may lead to the formation of such substances in the marine environment;
— Acid and alkaline compounds of such composition and in such quantities that they may seriously impair the quality of the marine environment;
— Substances which though not producing toxic effects, may become harmful because of the concentrations or quantities in which they are discharged, or which are liable to reduce amenities seriously or to endanger human life or marine organisms or to impair other legitimate uses of the sea;

— Pathogenic micro-organisms which are or may become harmful because of the concentrations and quantities in which they are discharged or which are liable to endanger human life or marine organisms, or impair other legitimate uses of the marine environment and the coastal waters in particular.

Annex III

Monitoring and Data Management

Monitoring

In the Protection of the marine environment against pollution from land based sources, monitoring can be defined as the measurement of the pollutant or its effects either man or elements or the marine resources for purposes of assessing and controlling exposure to the pollutant. Thus monitoring is used first to assess the d for pollution prevention measures and subsequently effectiveness of any protection measures introduced. If monitoring is to meet these objectives and be cost-effective it must be carefully designed and implemented.

Resources to be protected

One of the first thing to ascertain is what resources need protecting in the area concerned and the various pollutant sources and ways in which each could possibly be threatened. For example, the well-being of a nature, reserve, fish hatchery of fish resource might be threatened by the variety of substances. Similarly, the suitability of fish or shell-fish for human consumption might be affected by other substances such as mercury of arsenic which may adversely affect man, whilst not affecting fisheries.

Information on inputs

It is also important at an early stage to establish for each area the activities already practised and the pollutants likely to reach the sea via point, non-point and riverine sources.

A knowledge of the resources to be protected and which pollutants are most likely to affect them will allow attention to be focussed on those substances which appear most likely to be of concern, thereby reducing the amount of effort devoted to establishing a data base, on inputs. Information on inputs can also be used to focus environmental monitoring efforts on those pollutants most likely to be encountered in each area. If possible the scale of input should also be established, at least in order of magnitude terms. This will normally be fairly easy but in more accurate qualification will require improvements in the quality of data on both concentration and flow.

Information on inputs from direct discharges may be determined from descriptions of unit processes in use. If permit programmes have been established, information on controlled pollutants should be available from the permitting authority. Inputs from non-point sources are generally estimated by employing accepted formulae describing land use in the watershed and the associated run-off. In estimating point and non-point sources inputs, the pollutants of concern may include a broad range of substance for example, toxicants and nutrients.

Establishing baseline concentrations

Having decided what needs to be monitored, on the basis of what resources must be protected and which pollutants are likely to be of interest, the concentrations actually present in the environment can be established. This information can then be used to assess those protection measures necessary and/or their effectiveness. The need for control measures my be judged either by comparison of the concentrations found with some form of water quality criteria, for example maximum permissible concentration, or with similar data from other areas known not to be contaminated.

When baseline concentrations are being ascertained, the most appropriate substrate should be selected. Three options exist: water, biota and sediments, only rarely should it be necessary to analyse samples of all three. The choice will depend on the pollutant concerned, the water quality criteria selected and the nature of the pathways exposed. For example, water would be most suitable for nutrients, biochemical oxygen demand (BOD), pH and certain metals, but biota would be more appropriate for polychlorinated biphenyles (PCBs) or mercury, and undisturbed sediments can be particularly useful in time or spatial trend assessments.

Ongoing monitoring

Monitoring will be required to establish the effectiveness of pollution protection measures. Even if no reductions in inputs are deemed necessary it may be desired to check that the situation does not deteriorate. Whatever their purpose, monitoring programmes should designed to consider the receiving capacity of the environment as well as inputs. This means considering present water quality in relation to the desired quality, the scale of environmental protection measures taken in relation to the existing concentrations, nature of the pollutants present, the scale of their input and their removal processes. On this basis it will be possible to define what should be monitored and with what frequency.

Sampling and analysis

The number of samples collected and their nature should represent the substrate being monitored. Water quality, biological tissues and sediments can all be very variable even over short distances and the sampling strategy should, when necessary be tested statistically to ensure it is sound. The programme design should take account of the hydrographic characteristics of the area so as to avoid sampling the same body of water at different places as it moves under the influence of a current. Finally, the sample collected must be adjusted to the form in which the pollutant occurs in the environment or in the discharge streams.

Once a suitable sampling programme has been designed, it may be possible to bulk samples for analysis in order to reduce the analytical workload and costs. This will inevitably lead to the loss of some information and should only be considered if the complexity of the analytical technique demands it and/or the loss of information can be tolerated or if the monitoring is only to be used to pick up abnormalities such as in compliance monitoring.

Resource monitoring

In addition to monitoring the pollutants of interest in the selected substrate, it is essential that the state of the resource(s) be monitored. However, if adverse changes do occur it should not be assumed the protection measures taken were inadequate. For example, fish stocks decline due to fishing effort as well as pollution and undesirable plankton blooms occur for reasons other than nutrient enrichment. Biological effects monitoring is desirable but very few techniques can be applied routinely on a wide scale and most give unspecific responses. Once suitable effects monitoring techniques are available, they may be more attractive alternatives than purely chemical monitoring in environmental matrices.

Data and data management

Before the data from any monitoring programme are used, it is important that confidence limits be established and reported in order to ensure that the confidence with which recorded numbers are handled and interpreted is not misplaced. It is also necessary to decide how the data should be handled for future reference and use.

Limitations in the data and the extent to which they can be tolerated

The results obtained from any monitoring programme will be subject to errors of accuracy and precision, the size of which must be quantified. If precision is high and accuracy poor then all results for a set of analyses of the same sample will be very close together, for example, differing by no more than one per cent though they may differ from the true result by much more, possibly by as much as an order of magnitude. Some errors will derive from the nature of the samples. There can be minimized by proper statistical design of the sampling procedures and attention to the collection of uncontaminated samples.

All analytical procedures have inherent errors in precision and accuracy. To a greater or lesser extent either or both types of error can be compounded by operator or laboratory errors, which are often not recognized. However, by use of good analytical equipment and methods and by following a rigorous analytical quality assurance scheme, it should be possible to achieve high accuracy and precision for all analytical-data, and allow qualification of the scale errors.

Intercomparability requirements

In most cases where monitoring programmes are operated on a multilateral basis it is essential that the results obtained by all contributors are truly comparable.

Establishing comparable monitoring programmes may prove difficult. However, it is desirable that targets be set for comparability of the data.

Analytical comparability is only one aspect of monitoring data. The actual programmes run by different countries must also be comparable. It obviously will not be possible to compare results from three countries if one analyses water, another a fish species and another sediments. Even when agreement is reached on whether to sample water, biota or sediments it will be necessary to agree for example, which species of fish should be used, whether the water should be filtered before analysis or whether whole sediment should be analysed or only a particular size fraction.

Requirements for analytical quality control

It may be impossible to arrange that all contributors use identical analytical procedures. Even if they do, for the reasons given previously, intercomparability is not guaranteed. To establish whether differences do exist and to minimize them, a programme of intercalibration is essential. Each laboratory should assure the quality of its data by participating in intercalibration exercises and analysing at intervals reference materials containing certified concentrations of the pollutants of interest in appropriate matrices and concentrations.

Data storage, retrieval and exchange

Depending on the scale of the monitoring programme various methods of data storage and transfer may be appropriate. It is essential that the design of the storage/ retrieval system be carefully worked out to reflect the end use of the data both in its raw and interpreted form. The most efficient method in many respects is to use a computer. It is essential that the limitations of any set of data be instantly recognized when it is retrieved. To this end, information such as performance in a recognized interealibration exercise, analysis of reference material, etc. should be retrievable with the data. Ideally the data ,should be freely accessible by all contributors and the ,scientific community in general.

10 Male Declaration on Global Warming and Sea Level Rise (1989)

Environmental issues facing the world today clearly demonstrate that organisms and all the elements of nature including land, water and air cannot be exploited without far-reaching implications for the earth and its environmentally has been proved more conclusively than at any other time in history that the welfare of mankind is inextricably linked to the state of the environment. It has also been established that atmospheric concentrations of carbon dioxide and other greenhouse gases have increased over the last two centuries. These increases are seriously threatening to cause climate change, global warming and sea level rise, which have become common concerns of mankind.

There is now a broad scientific consensus that the global mean temperature could rise approximately by 1 to 2°C by the year 2020. It is predicted that even if the increases in atmospheric concentrations of greenhouse gases were to be brought to a standstill immediately, however unrealistic that may be, the global temperature and sea level wild continue to rise for decades to come.

Although the entire world would be adversely affected by these processes, low-laying, small, coastal and island States will face a decidedly greater predicament. Sea-level rise would cause extensive damage to the land

and infrastructure of those countries and even threaten the very survival of some island states. The possibility also exists of an increase in the frequency and/or intensity of natural disasters related to climate change, global warming and sea level rise.

Paradoxically the catalyst in this disturbing state of global environment has been the rapid development industrialization that was intended to lead to material ss. In view of the fact that the overloading of the atmosphere with greenhouse gases occurred primarily rough the actions of the industrialized nations during past two hundred years, these nations now have a moral obligation to initiate on an urgent basis, international action to stabilize and subsequently reduce emissions of greenhouse gases and to sponsor, as a matter of priority, an urgent worldwide programme of action to combat the serious implications of climate change, global warming and sea level rise. In addition, resources and technology should be made available by the industrialized nations, particularly to the most vulnerable States, which may not have the financial and technical means to address these problems.

A continuing dialogue between the small States and rest of the world on the issue of sea level rise needs to be initiated. The small States call for an international response especially from the developed and industrialized of the world. The likely effects of sea level rise urgently need to be established more accurately, and an effective international strategy for the small States of the world to cope with those impacts should be agreed upon a separate issue within a global strategy. In this connection, the WMO/UNEP Intergovernmental Panel Climate Change (IPCC) is recognized as the main m for the on-going work on science, impacts and response strategies of climate change.

In the light of the scientific consensus regarding the likelihood of climate change and global warming and ply concerned over the changing global environment its possible adverse effects, particularly the threat sea level rise, the Small States gathered here in Male from 14-18 November 1989, declare their intent to world collaborate and seek international co-operation to protect the low-lying small coastal and island States of the world from the dangers posed climate change, global warming and sea rise.

Therefore, we, The Representatives of the Small States Gathered here:

1. Decide to develop a Programme of action within the small States, for cooperation and exchange of information on strategies and policies in relation to climate change, global warming and sea level rise which are common concerns of mankind; and in particular, to:
   (a) establish in Action Group, initially comprising of representatives from the Caribbean, South Pacific, Mediterranean and the Indian Ocean regions, to oversee the implementation of the decisions and recommendations of the Small States Conference on Sea Level Rise, to coordinate a joint approach on the issues of climate-change, global warming and sea level rise, and to pursue and follow up on global and regional response strategies;
   (b) consider the establishment of a climate and sea level programme and a monitoring network as an important component within the global measuring systems: recognizing the urgent necessity to take initial measures to create a monitoring infrastructure, bearing in mind the specific interests of small developing island

States, to apply to the appropriate United Nations Agencies (in particular WMO, UNEP, UNESCO) for assistance in its implementation;

(c) mount a campaign to increase awareness of the international community of the particular vulnerability of the small States to sea level rise;

(d) consider the most effective manner in which the small States can participate in the work of the Intergovernmental Panel on Climate Change, and seek assistance for such participation; and

(e) seek assistance from the UN, its Agencies and other appropriate institutions in the implementation of the decisions contained in this Declaration.

Call upon all States of the world family of nations to take immediate and effective measures according to their capabilities and the means at their disposal, to control, limit or reduce the emission of greenhouse gases, and to consider ways and means of protecting the small States of the world which are most vulnerable to sea level rise.

Urge all States to take immediate measures to enhance energy efficiency and to formulate plans and strategies for a change over, as far as possible, to alternative, less environmentally harmful sources of energy.

Recommend that where necessary all States take Immediate measures to- establish the institutional framework to protect and manage their coastal zones and b~ enact legislation to facilitate such measures.

Call upon all States to undertake environmental impact assessment studies for all development projects, review existing development programmes in terms of environmental impact assessment and strengthen environmental management capabilities.

Recommend that small coastal and island States take adequate measures to maintain their aquifers and protect vulnerable natural ecosystems such as coral reefs and mangroves, which may already be at risk, as they can provide natural protection against the adverse effects of climate change, global warming and sea level rise.

Appeal to all States to embark on intensive afforestation and or revegetation programmes with emphasis on the selection of plants and trees suitable for the different soil conditions, and salt-tolerant varieties for the protection of coastal areas.

Recommend that research be intensified in understanding the complex interrelationships concerning climate change, greenhouse effect, sea level rise and their implications on the environment and also to determine methods of ameliorating the impacts of these changes on coastal ecosystems.

Urge the industrialized nations to develop modalities and mechanisms to facilitate funding, technology transfer and training in areas related to the causes and problems associated with the rise in sea level. In, this regard, the States facing immediate threat should be assigned a higher priority for assistance.

Support the call by the developing countries of the world for the strengthening of the existing funding, technology transfer and information mechanisms, not excluding the development of new mechanisms to assist them in implementing measures to control, limit or reduce emissions of greenhouse gases and adapt to and protect themselves from the adverse effects of unavoidable climate change, global warming and sea level rise.

Such mechanisms would also help to ensure that the transition to a more and environmentally sound worldwide programme of sustainable development can be achieved.

Call for negotiations for a framework convention on climate change to start as soon as possible after the adoption of the interim report of the Intergovernmental Panel on Climate Change.

11 International Convention on Oil Pollution Preparedness, Response and Cooperation (1990)

The Parties to the Present Convention,

Conscious of the need to preserve the human environment in general and the marine environment in particular.

Recognizing the serious threat posed to the marine environment by oil pollution incidents involving ships, offshore units, sea ports and oil handling facilities.

Mindful of the importance of precautionary measures and prevention in avoiding oil pollution in the first instance, and the need for strict application of existing international instruments dealing with maritime safety and marine pollution prevention, particularly the International Convention of the Safety of Life at Sea, 1974, as amended, and the International Convention for the Prevention of Pollution from ships, 1973, as modified by the Protocol of 1978 relating thereto, as amended, and also the speedy development of enhanced standards for the design, operation and maintenance of ships carrying oil, and of offshore units.

Mindful also that, in the event of an pollution incident, prompt and effective action is essential in order to minimize the damage which may result from such an incident.

Emphasizing the importance of effective preparation for combating oil pollution incidents and the important role which the oil and shipping industries have in this regard.

Recognizing further the importance of mutual assistance and international co-operation relating to matters including the exchange of information respecting the capabilities of States to respond to oil pollution incidents, the preparation of oil pollution contingency plans, the exchange of reports of incidents of significance which may affect the marine environment of the coastline and related interests of States, and research and development respecting means of combating oil pollution in the marine environment.

Taking account of the "polluter pays" principle as a general principle of international environmental law.

Taking account also of the importance of international instruments on liability and compensation for oil pollution damage, including the 1969 International Convention on Civil Liability for Oil Pollution Damage (CLC); and the 1971 International Convention on the Establishment of an International Fund for Compensation for Oil Pollution Damage (FUND); and the compelling need for early entry into force of the 1984 Protocols to the CLC and FUND Conventions.

Taking account further of the importance of bilateral and multilateral agreements and arrangements including regional conventions and agreements.

Bearing in mind the relevant provisions of the United Nations Convention on the Law of the Sea, in particular of its part XII.

Being aware of the need to promote international cooperation and to enhance existing national, regional and global capabilities concerning oil pollution preparedness and response, taking into account the special needs of the developing countries and particularly small island States.

Considering that these objectives may best be achieved by the conclusion of an International Convention on Oil Pollution Preparedness, Response and Co-operation.

Have agreed as follows:

Article 1

General Provisions

(1) Parties undertake, individually or jointly, to take all appropriate measures in accordance with the provisions of this Convention and the Annex thereto to prepare for and respond to an oil pollution incident.

(2) The Annex to this Convention shall constitute an integral part of the Convention and a reference to this Convention constitutes at the same time a reference to the Annex.

(3) This Convention shall not apply to any warship, naval or other ship owned or operated by a State and used, for the time being, only on government non-commercial service. However, each Party shall ensure by the adoption of appropriate measures not impairing the operations or operational capabilities of such ships owned or operated by it, that such ships act in a manner consistent so far as is reasonable and practicable, with this Convention.

Article 2

Definitions

For the purposes of this Convention:

(1) "Oil" means petroleum in any form including crude oil, fuel oil, sludge, oil refuse and refined products.

(2) "Oil pollution incident" means an occurrence or series of occurrences having the same origin, which results or many result in discharge of oil and which poses or may pose threat to the marine environment, or to the coastline or related interests of one or more States, and which requires emergency action or other immediate response.

(3) "Ship" means a vessel of any type whatsoever operating in the marine environment and includes hydrofoil boats, air-cushion vehicles, submersibles, and floating craft of any type.

(4) "Offshore unit" means any fixed or floating offshore installation or structure engaged in gas or oil exploration, exploitation, or production activities, or loading or unloading facilities.

(5) "Sea ports and oil handling facilities" means those facilities which present a risk of an oil pollution incident and includes, inter alia, sea ports, oil terminals, pipelines and other oil handling facilities.

(6) "Organization" means the International Maritime Organization.

(7) "Secretary-General" means the Secretary-General of the Organization.

Article 3

Oil Pollution Emergency Plans

(1) (a) Each Party shall require that ships entitled to fly its flag have on board a shipboard oil pollution

emergency plan as required by and in accordance with the provision adopted by the Organizations for this purpose.

(b) A ship required to have on board an oil pollution emergency plan in accordance with sub-paragraph (a) is subject, while in a port or at an offshore terminal under the jurisdiction of a Party, to inspection by officers duly authorized by that Party, in accordance with the practices provided for in existing international agreements or its national legislation.

(2) Each Party shall require that operators of offshore units under its jurisdiction have oil pollution emergency plans, which are co-ordinated with the national system established in accordance with article 6 and approved in accordance with procedures established by the competent national authority.

(3) Each Party shall require that authorities or operators in charge of such sea ports and oil handling facilities under its jurisdiction as it deems appropriate have oil pollution emergency plans or similar arrangements which are co-ordinated with the national system established in accordance with article 6 and approved in accordance with procedures established by the competent national authority.

Article 4

Oil Pollution Reporting Procedures

(1) Each Party shall:

(a) require masters or other persons having charge of ships flying its flag and person having charge of offshore units under its jurisdiction to report without delay any event on their ship or offshore unit involving a discharge or probable discharge of oil:

(i) in the case of a ship, to the nearest coastal State;

(ii) in the case of an offshore unit, to the coastal State to whose jurisdiction the unit is subject;

(b) require masters or other persons having charge of ships flying its flag and persons having charge of offshore units under its jurisdiction to report without delay any observed event at sea involving a discharge of oil or the presence of oil;

(i) in the case of a ship, to the nearest coastal State;

(ii) in the case of an offshore unit, to the coastal State to whose jurisdiction the unit is subject;

(c) require persons having charge of a sea ports and oil handling facilities under its jurisdiction to report without delay and event involving a discharge or a probable discharge of oil or the presence of oil to the competent national authority.

(d) instruct its maritime inspection vessels or aircraft and other appropriate services or officials to report without delay any observed event at sea or at a sea port or oil handling facility involving a discharge of oil or the presence of oil to the competent national authority or, as the case may be, to the nearest coastal State;

(e) request the pilots of civil aircraft to report without delay any observed event at sea involving a discharge of oil or the presence of oil to the nearest coastal State.

(2) Reports under paragraph (1)(a)(i) shall be made in accordance with the requirements developed by the Organization and based on the guidelines and general principles adopted by the Organization. Reports under paragraph (1)(a)(ii), (b),(c) and (d) shall be made in accordance with the guidelines and general principles adopted by the Organization to the extent applicable.

Article 5

Action on Receiving an Oil Pollution Report

(1) whenever a Party receives a report referred to in article 4 or pollution information provided by other sources, it shall:

(a) assess the event to determine whether it is an oil Pollution incident:

(b) assess the nature, extent and possible consequences of the oil pollution incident; and

(c) then, without delay, inform all States whose interests are affected or likely to be affected by such oil pollution incident, together with

(i) details of its assessments and any action it has taken, or intends to take, to deal with the incident, and

(ii) further information as appropriate, until the action taken to respond to the incident has been concluded or until joint action has been decided by such States.

(2) When the severity of such oil pollution incident so justifies, the Party shall provide the Organization directly or, as appropriate, through the relevant

regional organization or arrangements with the information referred to in paragraph (1) (b) and (c).

(3) When the severity of such oil pollution incident so justifies, other States affected by it are urged to inform the Organization directly or, as appropriate, through the relevant regional organizations or arrangements or their assessment of the extent of the threat to their interests and any action taken or intended.

(4) Parties should use, in so far as practicable, the oil pollution reporting system developed by the Organization when exchanging information and communicating with other States and with the Organization.

Article 6

National and Regional Systems for Preparedness and Response

(1) Each Party shall establish a national system for responding promptly and effectively to oil pollution incidents. This system shall include as a minimum:

(a) the designation of:

(i) the competent national authority or authorities with responsibility for oil pollution preparedness and response;

(ii) the national operational contact point or points, which shall be responsible for the receipt and transmission of oil pollution reports referred to in article 4; and

(iii) an authority which is entitled to act on behalf of the State to request assistance or to decide to render the assistance requested;

(b) a national contingency plan for preparedness and response which includes the organizational relationship of the various bodies involved, whether public or private, taking into account guidelines developed by the Organization.

(2) In addition, each Party, within its capabilities either individually or through bilateral or multilateral co-operation and, as appropriate, in co-operation with the oil and shipping industries, port authorities and other relevant entities, shall establish:

(a) a minimum level of pre-positioned oil spill combating equipment, commensurate with the risk involved, and programmes for its use;

(b) a programme of exercise for oil pollution response organizations and training of relevant, personnel;

(c) detailed plans and communication capabilities for responding to an oil pollution incident. Such capabilities should be continuously available; and

(d) a mechanism or arrangement to co-ordinate the response to an oil pollution incident with, if appropriate, the capabilities to mobilize the necessary resources.

(3) Each Party shall ensure that current information is provided to the Organization, directly or through the relevant regional organization arrangements, concerning;

(a) the location, telecommunication data and, if applicable, areas of responsibility of authorities and entities referred to in paragraph (1) (a);

(b) information concerning pollution response equipment and expertise in disciplines related to oil pollution response and marine salvage which may be made available to other States, upon request; and

(c) its national contingency plan.

Article 7

International Co-operation in Pollution Response

(1) Parties agree that, subject to their capabilities and type availability of relevant resources, they will co-operate and provide advisory services, technical support and equipment for the purpose of responding to an oil pollution incident, when the severity of such incident so justifies, upon the request of any Party affected or likely to be affected.

(2) A Party which has requested assistance may ask the Organization to assist in identifying sources of provisional, financing of the costs referred to in paragraph (1)

(3) In accordance with applicable international agreements, each Party shall take necessary legal or administrative measures to facilitate;

(a) the arrival and utilization in and departure from its territory of ships, aircraft and other modes of transport engaged in responding to an oil pollution incident or transporting personnel, cargoes, material and equipment required to deal with such and incident; and

(b) the expeditious movement into, through, and out of its territory of personnel, cargoes, materials and equipment referred to in sub-paragraph (a).

Article 8

Research and Developmen

(1) Parties agree to co-operate directly or, as appropriate, through the Organization or relevant regional organizations or arrangements in the promotion and exchange of results of research and

development programmes relating to the enhancement of the state-of-the-art of oil pollution preparedness and response, including technologies and techniques for surveillance, containment, recovery, dispersion, clean-up and otherwise minimizing or mitigating the effects of oil pollution, and for restoration.

(2) To this end, Parties undertake to establish directly or, as appropriate, through the Organization or relevant regional organizations or arrangements, the necessary links between Parties' research institutions.

(3) Parties agree to co-operate directly or through the Organization or relevant regional organizations or arrangements to promote, as appropriate, the holding on a regular basis of international symposia on relevant subjects, including technological advances in oil pollution combating techniques and equipment.

(4) Parties agree to encourage, through the Organization or other competent international organization, the development of standards for compatible oil pollution combating techniques and equipment.

Article 9

Technical Co-operation

(1) Parties undertake directly or through the Organization and other international bodies, as appropriate, in respect of oil pollution preparedness and response, to provide support for those Parties which request technical assistance;

(a) to train personnel;

(b) to ensure the availability of relevant technology, equipment and facilities;

(c) to facilitate other measures and arrangements to prepare for and respond to oil pollution incidents; and

(d) to initiate joint research and development programmes.

(2) Parties undertake to co-operate actively, subject to their national laws, regulations and policies, in the transfer of technology in respect of oil pollution preparedness and response.

Article 10

Promotion of Bilateral and Multilateral Co-operations in Preparedness and Response

Parties shall endeavour to conclude bilateral or multilateral for oil pollution preparedness and response. Copies of such agreements shall be communicated to the Organization which should make them available on request to Parties.

Article 11

Relation to other convention and international agreements

Nothing in this Convention shall be construed as altering the rights or obligations of any Party under any other convention or international agreement.

Article 12

Institutional Arrangements

(1) Parties designate the Organization, subject to its agreement and the availability of adequate resources to sustain the activity, to perform the following functions and activities;

(a) information services:

(i) to receive, collate and disseminate on request the information provided by Parties (see, for example, articles 5(2) and (3), 6(3) and 10) and relevant information provided by other sources; and

(ii) to provide assistance in identifying sources of provisional financing of costs (see, for example, article 7(2);

(b) education and training:

(i) to promote training in the field of oil pollution preparedness and response (see, for example, article 9); and

(ii) to promote the holding of international symposia (see, for example, article 8 (3));

(c) technical services;

(i) to facilitate to co-operation in research and development (see, for example, article 8)(1), (2) and (4) and 9(1)(d));

(ii) to provide advice to States establishing national or regional response capabilities; and

(iii) to analyse the information provided by Parties (see, for example, articles 5(2) and (3), 60 and 8(1) and relevant information provided by other sources and provide advice or information to States;

(d) technical assistance:

(i) to facilitate the provision of technical assistance to States establishing national or regional response capabilities; and

(ii) to facilitate the provision of technical assistance and advice, upon the request of States faced with major oil pollution incidents.

(2) In carrying out the activities specified in this article, the Organization shall endeavour to strengthen the ability of States .individually or through regional arrangements to prepare for and combat oil pollution incidents, drawing upon the experience of States, regional agreements and industry arrangements and paying particular attention to the needs of developing countries.

(3) The provisions of this article shall be implemented in accordance with a programme developed and kept under review by the Organization.

Article 13

Evaluation of the Convention

Parties shall evaluate within the Organization the effectiveness of the Convention in the light of its objectives, particularly with respect to the principles underlying co-operation and assistance.

Article 14

Amendments

(1) This Convention may be amended by one of the procedures specified in the following paragraphs.

(2) Amendment after consideration by the Organization:

(a) Any amendment proposed by a Party to the Convention shall be submitted to the Organization and circulated by the Secretary-General to all Members of the Organization and all Parties at least six months prior to its consideration.

(b) Any amendment proposed and circulated as above shall be submitted to the Marine Environment Protection Committee of the Organization for consideration.

(c) Parties to the Convention, whether or not Members of the Organization, shall be entitled to participate in the proceedings of the Marine Environment Protection Committee.

(d) Amendments shall be adopted by a two-thirds majority of only the Parties to the Convention present and voting.

(e) If adopted in accordance with sub-paragraph (d), amendments shall be communicated by the Secretary-General to all Parties to the Convention for acceptance.

(f) (i) An amendment to an article or the Annex of the Convention shall be deemed to have been accepted on the date on which it is accepted by two thirds of the Parties.

(ii) An amendment to an appendix shall be deemed to have been accepted at the end of a period to be determined by the Marine Environment Protection Committee at the time of its adoption, which period shall not be less than ten months, unless within that period an objection is communicated to the Secretary General by not less than one third of the Parties.

(g) (i) An amendment to an article or the Annex of the Convention accepted in conformity with sub-paragraph (f)(i) shall enter into force six months after the date on which it is deemed to have been accepted with respect to the Parties which have notified the Secretary-General that they have accepted it.

(ii) An amendment to an appendix accepted in conformity with sub-paragraph (f)(ii) shall enter into force six months after the date on which it is deemed to have been accepted with respect to all Parties with the exception of those which, before that date, have objected to it. A Party may at any time withdraw a previously communicated objection by submitting a notification to that effect to the Secretary-General.

(3) Amendment by a Conference:

(a) Upon the request of a Party, concurred with by at least one third of the Parties, the Secretary-General shall convene a Conference of Parties to the Convention to consider amendments to the Convention.

(b) An amendment adopted by such a Conference by a two-thirds majority of those Parties present and voting shall be communicated by the Secretary-General to all Parties for their acceptance.

(c) Unless the Conference decides otherwise, the amendment shall be deemed to have been accepted and shall enter into force in accordance with the procedures specified in paragraph (2)(f) and (g).

(4) The adoption and entry into force of an amendment constituting an addition of an Annex or an appendix shall be subject to the procedure applicable to an amendment to the Annex.

(5) Any Party which has not accepted an amendment to an article or the Annex under paragraph (2)(f)(i) or an amendment constituting an addition of an Annex or an appendix under paragraph (4) or has communicated an objection to an amendment to an appendix under paragraph (2)(f)(ii) shall be treated as a non-Party only for the purpose of the

application of such amendment. Such treatment shall terminate upon the submission of a notification of acceptance under paragraph (2)(f) (i) or withdrawal of the objection under paragraph (2) (g) (ii).

(6) The Secretary-General shall inform all Parties of any amendment which enters into force under this article, together with the date on which the amendment enters into force.

(7) Any notification of acceptance of, objection to, or withdrawal of objection to, an amendment under this article shall be communicated in writing to the Secretary-General who shall inform Parties of such notification and the date of its receipt.

(8) An appendix to the Convention shall contain only provisions of a technical nature.

Article 15

Signature, Ratification, Acceptance, Approval and Accession

(1) This Convention shall remain open for signature at the Headquarters of the Organization from 30 November 1990 until 29 November 1991 and shall thereafter remain open for accession. Any State may become Party to this Convention by:

(a) signature without reservation as to ratification, acceptance or approval; or

(b) signature subject to ratification, acceptance or approval, followed by ratification, acceptance or approval; or

(c) accession

(2) Ratification, acceptance, approval or accession shall be effected by the deposit of an instrument to that effect with the Secretary-General.

Article 16

Entry into Force

(1) This Convention shall enter into force twelve months after the date on which not less than fifteen States have either signed it without reservation as to ratification, acceptance or approval or have deposited the requisite instruments of ratification, acceptance, approval or accession in accordance with article 15.

(2) For States which have deposited an instrument of ratification, acceptance, approval or accession in respect of this Convention after the requirements for entry into force thereof have been met but prior to the date of entry into force, the ratification, acceptance, approval, or accession shall take effect on the date of entry into force of this Convention or three months after the date of deposit of the instrument, whichever is the later date.

(3) For States which have deposited an instrument of ratification, acceptance, approval or accession after the date on which this Convention entered into force, this Convention shall become effective three months after the date of deposit of the instrument.

(4) After the date on which an amendment to this Convention is deemed to have been accepted under article 14, any instrument of ratification, acceptance approval or accession deposited shall apply to this Convention as amended.

Article 17

Denunciation

(1) This Convention may be denounced by any Party at any time after the expiry of five years from the date on which this Convention enters into force for that Party.

(2) Denunciation shall be effected by notification in writing to the Secretary-General.

(3) A denunciation shall take effect twelve months after a receipt of the notification of denunciation by the Secretary-General or after the expiry of any longer period which may be indicated in the notification.

Article 18

Depositary

(1) This Convention shall be deposited with the Secretary-General.

(2) The Secretary-General shall:

(a) inform all States which have signed this Convention or acceded thereto of:

(i) each new signature or deposit of an instrument of ratification, acceptance, approval or accession, together with the date thereof; (ii) the date of entry into force of this Convention; and

(iii) the deposit of any instrument of denunciation of this Convention together with the date on which it was received and the date on which the denunciation takes effect;

(b) transmit certified true copies of this Convention to the Governments of all States which have signed this Convention or acceded thereto.

(3) As soon as this Convention enters into force, a certified true copy thereof shall be transmitted by the depositary to the Secretary-General of the United Nations for registration and publication in accordance with Article 102 of the Charter of the United Nations.

Article 19

Languages

This Convention is established in a single original in the Arabic, Chinese, English, French, Russian and Spanish languages, each text being equally authentic.

In Witness Whereof the undersigned, being duly authorized by their respective Governments for that purpose, have signed this Convention.

Done at London this thirtieth day of November one thousand nine hundred and ninety.

Annex: Reimbursement of Costs of Assistance

(1) (a) Unless an agreement concerning the financial arrangements governing action of Parties to deal with oil pollution incidents has been concluded on a bilateral or multilateral basis prior to the oil pollution incident. Parties shall bear the costs of their respective actions in dealing with pollution in accordance with sub-paragraph (i) or sub-paragraph (ii).

(i) If the action was taken by one Party at the express request of another Party, the requesting Party shall reimburse to the assisting Party the cost of its action. The requesting Party may cancel its request at any time, but in that case it shall bear the costs already incurred or committed by the assisting Party.

(ii) If the action was taken by a Party on its own initiative, this Party shall bear the cost of its action.

(b) The Principles laid down in sub-paragraph (a) shall apply unless the Parties concerned otherwise agree in any individual case.

(2) Unless otherwise agreed, the costs of action taken by a Party at the request of another Party shall be fairly calculated according to the law and current practice of the assisting Party concerning the reimbursement of such cots.

(3) The Party requesting assistance and the assisting Party shall, where appropriate, co-operate in concluding any action in response to a compensation claim. To that end, they shall, give due consideration to existing legal regimes. Where the action thus concluded does not permit full compensation for expenses incurred in the assistance operation the Party requesting assistance may ask the assisting Party to waive reimbursement of the expenses exceeding the sums compensated or to reduce the costs which have been calculated in accordance with paragraph (2), If may also request a postponement of the reimbursement of such costs. In considering such a request, assisting Parties shall give due consideration to the needs of the developing countries.

(4) The provisions of this Convention shall not be interpreted as in any way prejudicing the rights of Parties to recover from third parties the costs of actions to deal with pollution or the threat of pollution under other applicable provisions and rules of national and international law. Special attention shall be paid to the 1969 International Convention on Civil Liability for Oil Pollution Damage and the 1971 International Convention on the Establishment of an International Fund for Compensation for Oil Pollution Damage or any subsequent amendment to those Conventions.

INDEX